The Botanist's Companion

VOL. II

By William Salisbury
Cover Design by Alex Struik

Copyright © 2012 Alex Struik.

Alex Struik retains sole copyright to the cover design of this edition of this book.

All rights reserved. No part of this publication may be reproduced, stored in a retrieval system, or transmitted, in any form or by any means, electronic, mechanical, photocopying, recording or otherwise, without the prior permission of the copyright owner.

The right of Alex Struik to be identified as the author of the cover design of this work has been asserted in accordance with the Copyright, Designs and Patents Act 1988.

ISBN-13: 978-1480225251

ISBN-10: 1480225258

OR

AN INTRODUCTION TO THE KNOWLEDGE OF
PRACTICAL BOTANY, AND THE USES OF PLANTS.
EITHER GROWING WILD IN GREAT BRITAIN, OR
CULTIVATED FOR THE PUROSES OF
AGRICULTURE, MEDICINE, RURAL OECONOMY,
OR THE ARTS.

Contents

PREFACE TO THE SECOND VOLUME 7
PLANTS USEFUL IN AGRICULTURE 9
SECT. I.—GRASSES. 11
SECT. II.—ARTIFICIAL GRASSES 26
   HINTS AS TO THE LAYING DOWN LAND TO PERMANENT PASTURE. 37
SECT. III.—FODDER FROM LEAVES AND ROOTS. .. 42
SEC. IV.—GRAINS. 47
SECT. V.—MISCELLANEOUS ARTICLES. 57
PLANTS USEFUL IN THE ARTS. 61
SECT. VI.—BRITISH TREES AND SHRUBS. 61
SECT. VII.—PLANTS USEFUL IN MEDICINE. 76
SECT. VIII.—MEDICINAL PLANTS not contained in either of the BRITISH DISPENSATORIES. 132
   Observations on the Drying and Preserving of Herbs, &c. for Medicinal Purposes. 160
SECTION IX.—PLANTS USED FOR CULINARY PURPOSES 166
SECTION X.—CULINARY PLANTS NOT IN CULTIVATION. 185
SECTION XI.—PLANTS USEFUL IN DYEING. 196
SECTION XII.—-PLANTS USED IN RURAL OECONOMY. 212
SECTION XIII.—POISONOUS PLANTS GROWING IN GREAT BRITAIN. 214
   BITTER NAUSEOUS POISONS. 217
   ACRID POISONS. 224

> STUPEFYING POISONS. 227
>
> FOETID POISONS. 233
>
> DRASTIC POISONS 236
>
> POISONOUS FUNGI 239

SECTION XIV.—PLANTS NOXIOUS TO CATTLE ... 240

SECTION XV.—PLANTS NOXIOUS IN AGRICULTURE. 243

> Creeping-rooted Weeds. 246
>
> Perennial Weeds. 248

SECTION XVI.-EXOTIC TREES AND SHRUBS. 250

SECTION XVII.- FOREIGN HARDY HERBACEOUS PLANTS. 274

SECTION XVIII.-HARDY ANNUAL FLOWERS 295

SECTION XIX.-BIENNIAL FLOWERS. 298

SECTION XX.-TENDER ANNUAL FLOWERS. 299

SECTION XXI.-FOREIGN ALPINE PLANTS. 300

APPENDIX 309

> BRITISH PLANTS CULTIVATED FOR ORNAMENTAL PURPOSES 309
>
> MISCELLANEOUS ARTICLES 320
>
> OBSERVATIONS on the BLEEDING TREES, and procuring the Sap for making Wine, and brewing Ale. 329

"Behold I have given you every herb bearing seed, and every tree yielding fruit, and to you it shall be for meat."

PREFACE TO THE SECOND VOLUME

In demonstrating the Plants which occur in our annual herborizing excursions, I have found it necessary to put into the hands of my pupils some Manual of Botany; and in so doing I have found all that have yet been published, deficient in one or two essential points, and particularly as relating to the uses to which each plant is adapted; with out which, although the charms of the Flora are in themselves truly delightful, yet the real value of Botanic knowledge is lost. The study of plants, so far as regards their uses and culture, has engaged my particular attention for the last twenty-five years, during which time I had the honour of conducting a series of experiments on the growth of plants, for the Board of Agriculture, which gave me an opportunity of ascertaining many facts relative to our Grasses, &c. an account of which, I have had some time ready for publication. The necessity of a work of this kind in my present profession, has therefore induced me to abridge it and put it to press; as such I offer it to the Public. To the Subscribers to my Botanic Garden this will also prove of great service; it being intended to arrange the plants in their several departments, so as to make it a general work of reference both in the fields or garden. In the department which treats of the Vegetables used for medicinal purposes, I have given as ample descriptions as the nature of the work will admit of, having in view the very necessary obligation which the younger branch of the profession are under, of paying attention to the subject.

In prosecuting this work, I have been more actuated by a desire to render to my pupils and others, useful information, than that of commencing Author on such a subject; and writing for the press has been but very little my employment, I trust that an ample excuse will be granted for any errors that may appear, or for the want of that happiness of diction with which more able and accomplished Authors may be endowed.

BOTANIC GARDEN,

Sloane Street, May 1816.

PLANTS USEFUL IN AGRICULTURE.

OBSERVATIONS ON THE CULTURE OF GRASSES, AND ON SAVING SEEDS, &c.

It is now fifty years since the celebrated Stillingfleet observed, "that it was surprising to see how long mankind had neglected to make a proper advantage of plants, of so much importance to agriculture as the Grasses, which are in all countries the principal food of cattle." The farmer, for want of distinguishing and selecting the best kinds, fills his pastures either with weeds or improper plants, when by making a right choice he would not only procure a more abundant crop from his land, but have a produce more nourishing for his flock. One would therefore naturally wonder, after this truth has been so long published, and that in an age when agriculture and the arts have so much improved, that Select Seeds of this tribe of plants are scarcely to be produced.

From the experience I have had on this subject, I find their culture is attended with certain difficulties, which arise not so much from the nature of the plants, as from the labour requisite to this purpose, great attention being necessary for saving Grass-seeds at the seasons when the farmer must exert all the strength of his husbandmen to get his other business accomplished.

The only mode by which this can be effected is by selecting a proper soil for the kinds intended to be saved. The seeds should be drilled into the ground at about one foot distance; and care taken that the plants are duly weeded of all other kinds that may intrude themselves, before they get too firm possession of the soil. The hoe should be frequently passed between the drills, in order both to keep the land clean and to give vigour to the young plants. The sowing may be done either in the spring or in the month of September, which will enable the crop to go to seed the following spring. In order to preserve a succession of crops, it is necessary every season to keep the ground clean all the summer months, to dig or otherwise turn up the land between the drills early in the spring, and to be particular in the other operations until the seeds ripen. Now this business being so inconvenient to the farmer, it is not to be wondered at, that, wherever attempts of this kind have been made, they should fail from want of the necessary care as above stated, without which it is needless to speculate in such an undertaking. There is nevertheless still an opportunity, for any one who would give up his land and time to the pursuit, to reap a rich and important harvest; as nothing would pay him better, or redound more to his credit, than to get our markets regularly supplied with select seeds of the best indigenous Grasses, so that a proper portion of them may be used for forming pasture and meadow-land.

The above hints are not thrown out by a person who wishes to speculate in a theory which is new, but by one who has cultivated those plants himself both for seed and fodder, and who would readily wish to promote their culture by stating a mode which has proved to him a profitable pursuit, and for which he has, already, been honoured with a reward form the Society of Arts.

The following observations are intended to embrace such kinds only as are likely to be cultivated, with those that are distinguished for some particular good properties; as it would be impossible within the limits of this small memorandum to enumerate all the plants that are eaten by cattle. The same mode shall be pursued under all the different heads in this department.

SECT. I.—GRASSES.

1. ANTHOXANTHUM odoratum. SWEET-SCENTED VERNAL-GRASS.—This is found frequently in all our best meadows, to which it is of great benefit. It is an early, though not the most productive grass, and is much relished by all kinds of cattle. It is highly odoriferous; if bruised it communicates its agreeable scent to the fingers, and when dry perfumes the hay. It will grow in almost any soil or situation. About three pounds of seed should be sown with other grasses for an acre of land.

2. ALOPECURUS pratensis. MEADOW FOX-TAIL-GRASS.—One of our most productive plants of this tribe: it grows best in a moist soil, is very early, being often fit for the scythe by the middle of May. About two bushels of seed will sow an acre, with a proportionate quantity of Clover; which see.

3. ALOPECURUS geniculatus. FLOTE FOX-TAIL-GRASS.—Is very good in water meadows, being nutritive, and cattle in general are fond of it. We do not know if the cultivation of this plant has as yet been attempted.

4. AGROSTIS capillaris. FINE BENT-GRASS.—Dr. Walker, in his History of the Hebrides, speaks very favourably of this grass. I have therefore noticed it here, but I do not think it so good as many others. It grows on the sandy hills near Combe Wood in Surrey, and forms the principal part of the pasturage; but it is neither very productive, nor are cattle observed to thrive on it. The seeds are very small; one peck would sow an acre.

5. AGROSTIS pyramidalis. FIORIN-GRASS [Footnote: Fiorin is the Irish name of butter].—No plant has engaged

the attention of the farmer more than this grass, none ever produced more disputes, and none is perhaps so little understood. It is perfectly distinct from any species of Agrostis indigenous to this country: it is introduced by Dr. Richardson, and to that gentleman's extraordinary account of it we are indebted for numerous mistakes that have been made respecting it. It is an amphibious plant, thriving only in water or wet soils, is very productive, and the stalks after a summer's growth secrete a large quantity of sugar. It has the power, when the stalks are ripe, of resisting putrefaction, and will become blanched and more nutritious by being cut and laid in heaps in the winter season, at which time only it is useful. The cultivator of this plant must not expect to graze his land, but allow all the growth to be husbanded as above; and although it will not be found generally advantageous on this account, it nevertheless may be grown to very great advantage either in wet soils, or where land can be flooded at pleasure.

The seeds are often barren; and the only mode is to plant the shoots or strings in drills at nine inches apart, laying them lengthways along the drills, the ends of one touching the other.

6. AIRA aquatica. WATER HAIR-GRASS.—This is an aquatic, and very much relished by cattle, but cannot be propagated for fodder. Water-fowl are very fond of the young sweet shoots, as also of the seeds; it may therefore be introduced into decoys and other places with good effect. Pulling up the plants and throwing them into the water with a weight tied to them, is the best mode of introducing it.

7. ARUNDO arenaria. SEA-SIDE REED-GRASS.—This is also of no value as fodder, but it possesses the property of forming by its thick and wiry roots considerable hillocks

on the shores where it naturally grows: hence its value on all new embankments. If it be planted in a sandy place, during its growth in the summer the loose soil will be collected in the herbage, and the grass continues to grow and form roots in it; and thus is the hillock increased. Local acts of parliament have been passed, and now exist, for preventing its destruction on the sea-coast in some parts of Great Britain, on this account.

8. ARUNDO Phragmites. COMMON REED.—Is useful for thatching, and making slight fences; it grows best in ponds near streams of water; it does not often seed, but it could easily be introduced to such places by planting its roots in spring: it is a large-growing plant; and where herbage may be wanted either for beauty or shelter for water-fowl, nothing can be more suitable, and the reeds are of great value.

9. AVENA flavescens. YELLOW OAT-GRASS.—Is much eaten by cattle, and forms a good bottom. It has the property of throwing up flowerstalks all the summer; hence its produce is considerable, and it appears to be well adapted to pasture. The seeds of this grass are not to be obtained separately; hence it is not in cultivation. It is however worthy of attention, as the seeds are produced very abundantly in its native places of growth. It will grow either in wet or dry soils.

10. AVENA pubescens. ROUGH OAT-GRASS.—This appears to have some merits, but the foliage is extremely bitter. It grows in dry soils.

11. AVENA elatior. TALL OAT-GRASS.—From the good appearance of this grass some persons have recommended it as likely to be useful for forming meadows; but it is excessively bitter, and is not liked by cattle generally,

though when starved they are sometimes observed to eat of it. There is a variety of it with knobby roots which is found to be a most troublesome and noxious weed in arable lands, particularly in some parts of the coast of Hampshire where it abounds. This variety was some years ago introduced into the island of St. Kitts, and it has since taken such firm possession of the land as to render a large district quite useless. Persons should be cautious how they speculate with weeds from appearances only.

12. BRIZA media. QUAKING-GRASS.—Is common in meadow land, and helps to make a thick bottom; it does not however appear to be worth the trouble of select culture. It is bitter to the taste.

13. BROMUS mollis. SOFT BROME-GRASS.—Mr. Curtis has given a very clear account of this grass, which he says predominates much in the meadows near London, but that the seeds are usually ripe and the grass dried up before the hay time: hence it is lost; and he in consequence considered it only in the light of a weed. It has seldom occurred to me to differ in opinion from this gentleman, who certainly has given us, as far as it goes, a most perfect description of our useful grasses: but experience has convinced me that the Soft Brome-Grass, which seeds and springs up so early, makes the chief bulk of most of our meadows in March and April; and although it is ripe and over, or nearly so, by the hay harvest, yet the food it yields at this early season is of the greatest moment, as little else is found fit for the food of cattle before the meadow is shut up for hay, and this plant being eaten down at that season is not any loss to the hay crop. Whoever examines the seeds of this grass will be led to admire how wonderfully it is fitted to make its way into the soil at the season of its ripening, when the land is thus covered with the whole produce of a meadow. I notice this curious piece of

mechanism [Footnote: Many seeds of the grasses are provided with awns which curl up in dry weather and relax with moisture. Thus by change of atmosphere a continued motion is occasioned, which enables the seeds to find their way through the foliage to the soil, where it buries itself in a short time in a very curious manner.], not that it is altogether peculiar to this plant, but to show that Nature has provided it means of succeeding in burying itself in the ground, when all the endeavours of man could not sow the land with any other to answer a similar purpose. If the seeds of this grass were collected and introduced in some meadows where it is not common, I am sure the early feeding would be thereby improved.

The seeds are sometimes mixed with those of Rye-grass at market, and it is known by the name of Cocks: it has the effect of reducing such samples in value, but I should not hesitate in preferring such to any other. If any one should be inclined to make the above experiment, two pecks of the seed sown on an acre will be sufficient.—-See Treatise on Brit. Grasses by Mr. Curtis, edit. 5.

14. CYNOSURUS cristatus. CRESTED DOG'S-TAIL-GRASS.—A very fine herbage, and much relished by sheep, &c.; it grows best in fine upland loam, where it is found to be a most excellent plant both for grazing and hay. The seeds are to be purchased sometimes at the seedshops. About twelve pounds will sow an acre.—-See Observations on laying Land to Grass, in the Appendix to this work.

15. CYNOSURUS coeruleus. BLUE DOG'S-TAIL-GRASS.—Dr. Walker states this plant to be remarkably agreeable to cattle, and that it grows nearly three feet high in mountainous situations and very exposed places. As this grass does not grow wild in this part of the country, we have no opportunity of considering its merits. In our

Botanic Garden it seldom exceeds the height of ten inches or a foot.

It is the earliest grass of all our British species, being often in bloom in February.

The above intelligent gentleman, who seems to have studied the British Gramina to a considerable extent, says that the following kinds give considerable food to sheep and cattle in such situations; I shall therefore mention their names, as being with us of little esteem and similar to the above.

Phleum alpinum. Eriophorum polystachion. Festuca decumbens. Carex flavescens. Carex gigantea, probably Pseudocyperus. Carex trigona, probably vulpina. Carex elata, probably atrata. Carex nemorosa, probably pendula. And he is of opinion that the seeds may be sown to advantage. Be this as may, the observation can only apply to situations in the north of Britain, where he has seen them wild; in this part of the island we have a number of kinds much better adapted to soil, climate, and fodder.

16. DACTYLIS glomerata. ROUGH COCK'S-FOOT-GRASS.—Has a remarkable rough coarse foliage, and is of little account as a grass for the hay-stack; but from its early growth and great produce it is now found to be a useful plant, and is the only grass at this time known that will fill up the dearth experienced by graziers from the time turnips are over until the meadows are fit for grazing. Every sheep-farm should be provided with a due portion of this on the land; but no more should be grown than is wanted for early feed, and what can be kept closely eaten down all the season. If it is left to get up it forms large tufts, and renders the field unsightly, and scarcely any animal will eat it when

grown old or when dried in the form of hay. The seed is to be bought; two bushels per acres is sown usually alone.

17. FESTUCA elatior. TALL FESCUE-GRASS.—This in its wild state has been considered as a productive and nutritive grass; it grows best in moist places; but the seeds have been found in general abortive, and the grass consequently only to be propagated by planting the roots, a trouble by far too great to succeed to any extent.—See Poa aquatica.

18. FESTUCA duriuscula. HARD FESCUE-GRASS.—A very excellent grass both for green fodder and hay, and would be well worth cultivating; but the seeds have not hitherto been saved in any quantity.

I have seen a meadow near Bognor where it formed the principal part of the herbage; and it was represented to me by the owner as the best meadow in the neighbourhood, and the hay excellent [Footnote: Mr. Curtis observes that this grass grows thin on the ground after a time. I have sometimes observed this to be the case in the Botanic Garden, but it is otherwise in its native state of growth. Nothing stands the dry weather better, or makes a more firm sward.].

The seeds of this grass are small, and about one bushel would sow an acre of ground.

19. FESTUCA rubra. RED or CREEPING FESCUE-GRASS.—A fine grass, very like duriuscula; but it is not common in this part of the country; it grows plentifully on the mountains in Wales.

It does not produce fertile seeds with us in the garden.

20. FESTUCA pratensis. MEADOW FESCUE-GRASS.—No plant whatever deserves so much the attention of the graziers as this grass. It has been justly esteemed by Mr. Curtis and all other persons practically acquainted with the produce of our meadows. It will grow in almost any soil that is capable of sustaining a vegetable, from the banks of rivulets to the top of the thin-soiled calcareous hills, where it produces herbage equal to any other plant of the kind; and all descriptions of cattle eat it, and are nourished by the food. The plant is of easy culture, as it yields seeds very abundantly, and they grow very readily. I have made some excellent meadows with this seed, which after a trial of ten years are now equal to any in the kingdom. The culture of the seed selected is now nearly lost, which is a misfortune, I had almost ventured to say a disgrace, to our agriculture.

If the farmer could get his land fit for meadow laid down with one bushel of this seed, one bushel of Alopecurus pratensis, three pounds of Anthoxanthum, and a little Bromus mollis, with Clover, I will venture to predict experience will induce him to say, "I will seek no further."

21. FESTUCA ovina.—SHEEP'S FESCUE-GRASS.—This is very highly spoken of in all dissertations that have hitherto been written on the merits of our grasses; but its value must be confined to alpine situations, for its diminutive size added to its slow growth renders it in my opinion very inferior to the duriuscula. In fact, I am of opinion that these are often confounded together, and the merits of the former applied to this, although they are different in many respects. Those who wish to obtain more of its history may consult Stillingfleet's Observations on Grasses, p. 384.

22. FESTUCA vivipara. VIVIPAROUS FESCUE-GRASS.—This affords a striking instance of the protection

that Nature has contrived for keeping up the regular produce of the different species of plants; as when the Festuca ovina is found in very high mountainous situations, places not congenial to the ripening seeds of so light a nature, the panicle is found to become viviparous, i.e. producing perfect plants, which being beaten down with heavy rains in the autumn, readily strike root in the ground.

This plant was introduced into our garden many years ago, and still preserves this difference; otherwise it is in all respects the same as the Festuca ovina.

23. FESTUCA pinnata. SPIKED FESCUE-GRASS.—I have observed this near the Thames side to be the principal grass in some of the most abundant meadows; and as the seeds are very plentiful, I am of opinion it might be very easily propagated: it is, however, not in cultivation at present.

24. FESTUCA loliacea. DARNEL FESCUE-GRASS.— This in appearance is very like the Lolium perenne, but is a more lasting plant in the ground. Where I have seen it wild, it is certainly very good; but it is liable to the objection of Festuca elatior, the seeds grow but sparingly.

25. HOLCUS lanatus. YORKSHIRE GRASS, or MEADOW SOFT-GRASS.—This has been much recommended as fit for meadow-land. I am not an advocate for it. It is late in blooming, and consequently not fit for the scythe at the time other grasses are; and I find the lower foliage where it occurs in meadows to be generally yellow and in a state of decay, from its tendency to mat and lie prostrate. I hear it has been cultivated in Yorkshire; hence probably its name. Two bushels of the seed would sow an acre; and it is sometimes met with in our seed-shops. It will grow in any soil, but thrives best in a moist loam.

26. HOLCUS mollis. CREEPING SOFT-GRASS.—Mr. Curtis in the third edition of his Treatise on Grasses says, he is induced to have a better opinion than formerly of this grass, and that Mr. Dorset also thinks it may be cultivated to advantage in dry sandy soils. I have never seen it exhibit any appearance that has indicated any such thing, and do not recommend it.

27. HORDEUM pratense. MEADOW BARLEY-GRASS.—This is productive, and forms a good bottom in Battersea meadows: but although I have heard it highly recommended, I should fear it was much inferior to many others. One species of Barley-grass, which grows very commonly in our sea-marshes, the Hordeum maritimum, is apt to render cattle diseased in the mouth, from chewing the seeds, which are armed with a strong bristly awn not dissimilar to the spike of this grass.

28. LOLIUM perenne. RAY- or RYE-GRASS.—This has been long in cultivation, and is usually sown with clover under a crop of spring corn. It forms in the succeeding autumn a good stock of herbage, and the summer following it is commonly mown for hay, or the seed saved for market, after which the land is usually ploughed and fallowed, to clear it of weeds, or as a preparation for Wheat, by sowing a crop of Winter Tares or Turnips. The seed is about six or eight pecks per acre, and ten pounds of Clover mixt as the land best suits. Although this is a very advantageous culture for such purposes, and when the land is not to remain in constant pasture; yet it is by no means a fit grass for permanent meadow, as it exhausts the soil, and presently goes into a state of decay for want of nourishment, when other plants natural to the soil are apt to overpower it. There are several varieties of this grass. Some I have seen with the flowers double, others with branched panicles; some that grow very luxuriantly, and others that are little

better than annuals; and there is also a variety in cultivation called PACEY's Rye-grass, much sought for. But I am of opinion that nothing but a fine rich soil will produce a very good crop, and that the principal difference, after all, is owing more to cultivation or change of soil, than to any real difference in the plant itself.

29. MELICA coerulea. BLUE MELIC-GRASS.—This is common on all our heaths; it appears coarse, and not a grass likely to be useful. Yet this kind is spoken of by Dr. Walker under the name of Fly-bent, who says it is one of the most productive and best grasses for sheep-feed in the Highlands of Scotland, where it grows to the height of three feet, a size to which it never attains in this part of the country. It is found in all soils, both in dry and boggy places.

30. PANICUM germanicum. GERMAN PANIC, or MOHAR.—I notice this plant here, although it is not a native of this country; neither is it in cultivation. It was introduced some years since by Sir Thomas Tyrrwhit from Hungary. It is said there to be the best food of all others for horses; and I think it might be cultivated to advantage on high sandy soils, as a late crop of green fodder. The seeds are similar to Millet [Footnote: The Hungarian horses are remarked for their sleekness, and it is said that it is in consequence of being fed on Mohar.].

31. PANICUM crus galli. COCK'S-FOOT-PANIC-GRASS.—This plant has, I believe, never been recommended for cultivation; but it possesses qualities which render it worth attention: it will sometimes grow to the height of four feet, is very fine food for cattle, and will no doubt make excellent hay. It stands dry weather better than most other grasses I know. The seeds will not vegetate before May, and the crop not in perfection till late

September. In dry soils I think it could be cultivated to advantage if sown among a crop of Tares or Rye in the autumn; and after they are cut in summer, this would spring up and be a valuable acquisition in a dry autumn, as it would seldom fail producing an abundant crop.

It grows thick, and would tend to clear the land as a smothering crop over weeds: it is annual.

32. PHALARIS arundinacea. REED CANARY-GRASS.—This is not in cultivation, but grows plentyfully on the muddy banks of the Thames; it will also grow very well in a moderately dry soil; and I have observed that cattle eat it when it is young. As it is early and very productive, as well as extremely hardy, I think it might become valuable as early feed. The seeds of this plant do not readily grow, but it might easily be introduced by planting the roots in the spring. The Striped or Ribbon Grass of the flower garden is only a variety of this. See Poa aquatica.

33. PHLEUM pratense. TIMOTHY-GRASS, or MEADOW-CAT'S-TAIL-GRASS.—Is very coarse and late, and consequently not equal to many of our grasses either for hay or pasture. It has been highly recommended in America, where it may probably have been found to answer better than it has done with us in cultivation. The seed used to be imported from New York, and met with a ready sale; but I believe it is seldom imported at this time. Dr. Walker says the seeds were taken from South Carolina (where it was first cultivated) to that State, by one Timothy Hanson, from whence it acquired its name.

The same gentleman supposes it may be introduced into the Highlands of

Scotland with good effect, but is of my opinion as to its utility in England.—Rural Economy of the Hebrides, vol. ii. p. 27.

34. PHLEUM nodosum. BULBOUS CAT'S-TAIL-GRASS. (Phleum pratense var. ? Hudson.)—This affects a drier soil than the Timothy-grass: it grows very frequently in dry thin soils, where it maintains itself against the parching sun by its bulbous roots, which lie dormant for a considerable time, but grow again very readily when the wet weather sets in,—a curious circumstance, which gives us an ample proof of the wise contrivance of the great Author of Nature to fertilize all kinds of soil for the benefit of his creatures here below. There is another instance of this in the Poa bulbosa, Bulbous Meadow-grass, which grows on the Steine at Brighton, and which I have kept in papers two years out of ground, and it has vegetated afterwards.

35. POA annua. ANNUAL MEADOW-GRASS.—This is the most general plant in all nature: it grows in almost every situation where there is any vegetation. It has been spoken of as good in cultivation, and has had the term Suffolk grass applied to it, from its having been grown in that county. I have never seen it in such states, neither can I say I should anticipate much benefit to arise from a plant which is not only an annual, but very diminutive in size.

36. POA aquatica. WATER MEADOW-GRASS.—This is quite an aquatic, but is eaten when young by cattle, and is very useful in fenny countries: it is highly ornamental, and might be introduced into ponds for the same purpose as Arundo Phragmites: it might also be planted with Festuca elatior and Phalaris arundinacea, in wet dug out places,

where it would be useful as fodder, and form excellent shelter for game.

37. POA fluitans. FLOTE FESCUE-GRASS.—This would be of all others the most nutritive and best plant for feeding cattle; but it thrives only in water. I have noticed it only because it is highly recommended by the editor of Mr. Curtis's Observations on British Grasses, 5th edit. The cattle are very fond of it; but it is not to be cultivated, unless it be in ponds, being perfectly aquatic.

Linnaeus speaks of the seeds being collected and sold in Poland and Germany as a dainty for culinary purposes; but I have never seen it used here, neither are the seeds to be collected in great quantities. Stillingfleet, on the authority of a Mr. Dean, speaks highly of its merits in a water-meadow, and also quotes Mr Ray's account of the famous meadow at Orchiston near Salisbury. There this, as well as Poa trivialis, most certainly is in its highest perfection; but the real and general value of grasses or other plants must not be estimated by such very local instances, when our object is to direct the student to a general knowledge of the subject. See Curtis, art. Poa trivialis.

38. POA trivialis. ROUGH-STALKED MEADOW-GRASS.—Those who have observed this grass in our best watered meadows, and in other low pasture-land, have naturally been struck with its great produce and fine herbage. In some such places it undoubtedly appears to have every good quality that a plant of this nature can possess; it is a principal grass in the famous Orchiston meadow near Salisbury, and its amazing produce is mentioned in the Bath Agricultural Papers, vol. i. p. 94: but persons should not be altogether caught by such appearances; for I have seen it in some lands, and such as

would produce good red Clover, a very diminutive and insignificant plant indeed.

When persons wish to introduce it, they should carefully examine their neighbouring pastures, and see how it thrives in such places. The seeds are small, and six pounds would be sufficient for an acre, with others that affect a similar soil.

39. POA pratensis. SMOOTH-STALKED MEADOW-GRASS.—This is also a grass of considerable merit when it suits the soil; it affects a dry situation, and in some such places it is the principal herbage; but I have cultivated this by itself for seed in tolerably good land, and after some time I found it matted so much by its creeping roots as to become quite unproductive both of herbage and seed. Care should therefore be taken that only a proper portion of this be introduced. The seeds of this and Poa trivialis are the same in bulk, and probably the same proportion should be adopted. The seeds of both species hang together by a substance like to cobwebs, when thrashed, and require to be rubbed either in ashes or dry sand to separate them before sowing.

SECT. II.—ARTIFICIAL GRASSES

[Footnote: This technical term is generally known to farmers. It is applied to Clovers, and such plants as usually grow in pastures, and not strictly Gramina.].

Under this term are included such plants as are sown for fodder, either with a view to form permanent pastures when mixed with the grasses, or as intermediate crops on arable land. In those cases they are usually sown with a spring crop of Oats or Barley, and the artificial grasses are protected after the harvest by the stubble left on the ground, affording the succeeding season a valuable crop, either for pasturage or hay.

40. ACHILLEA Millefolium. YARROW.—This has been much recommended for sheep feed; but I observe it is frequently left untouched by them if other green herbage is found on the land. It will thrive in almost any soil, but succeeds best in good loam. The seed used is about twelve pounds per acre.

41. ANTHYLLIS vulneraria. KIDNEY VETCH.—This plant is not in cultivation, but it has been noticed that where it grows naturally the cows produce better milk and in greater quantity. It grows best in calcareous soils: the seeds are large, and easily collected. This plant well deserves attention.

42. CICHORIUM Intybus. CICHORY, or BLUE SUCCORY.-Much has been said of the good properties of this plant; and if it has them to the full extent mentioned by different authors, I wonder there is not little else than Cichory grown in this country. It is very prolific, and will grow extremely quick after the scythe during the summer

months: but I fear, from the observations I have made, that it does not possess the fattening quality it is said to have. The plant is so extremely bitter, that although cattle may be inclined to feed on it early in the spring, yet as the season advances and other herbage more palatable is to be met with, it is left with its beautiful blue flowers and broad foliage to rob the soil and adorn our fields, to the regret of the farmer. It grows wild in great abundance in Battersea fields, where my late friend Mr. Curtis used ludicrously to say that bad husbandry was exhibited to perfection. This plant is there continually seen in the greatest abundance, where the ground has not been lately disturbed, even under the noses of all the half-starved cattle of that neighbourhood that are turned in during the autumn.

The root dried and ground to a powder will improve Coffee, and is frequently drunk therewith, especially in Germany, where it is prepared in cakes and sold for that purpose.

43. HEDYSARUM Onobrychis. SAINT-FOIN.—This is certainly one of the most useful plants of this tribe, and in the south of England is the life and support of the upland farmer: in such places it is the principal fodder, both green and in hay, for all his stock. I have not observed it to be cultivated in Worcestershire or Herefordshire, where there appears to be much land that would grow it, and which is under much inferior crops. The seed sown is about four bushels per acre. A mistake is often made in mentioning this plant. The newspapers, in quoting prices from Mark Lane, call it Cinquefoil, a very different plant, (Potentilla) of rather a noxious quality. See Gleanings on Works of Agriculture and Gardening, p. 88, where a curious blunder occurs of this kind.

44. LATHYRUS pratensis. MEADOW VETCHLING.—Abounds much in our natural meadows, particularly in the best loamy soils, where it is very productive and nutritious. It is not in cultivation, for the seeds do not readily vegetate; a circumstance much to be regretted, but unfortunately the case with several of our other Tares, which would otherwise be a great acquisition to our graziers.

45. LOTUS corniculatus. BIRD'S-FOOT-LOTUS.—There are several varieties of this plant; one growing on very dry chalky soils, and which in such places helps to make a good turf, and is much relished by cattle. The other varieties grow in marshy land, and make much larger plants than the other. Here it is also much eaten; and I have also noticed it in hay, where it appears to be a good ingredient. As it thus appears to grow in any situation, there is no doubt, if the seeds were collected, that it might be cultivated with ease, and turn to good account in such land as is too light for Clover. In wet and boggy situations it becomes very hairy, and in this state its appearance is very different from that which it has when growing in chalk, where it is perfectly smooth.

This plant should not be overlooked by the experimental farmer.

It is very highly spoken of in Dr. Anderson's Essays on Agriculture, under the mistaken name of Astragalus glycophyllos, p. 489; but a truly practical account is given of it by Ellis in his Husbandry, p. 89, by the old name Lady-Finger-Grass.

46. MEDICAGO falcata. YELLOW MEDIC.—Is nearly allied to Lucerne, and is equally good for fodder; it will grow on land that is very dry, and hence is likely to become a most useful plant; its culture has, however, been tried but

partially. Some experiments were made with this plant by Thomas Le Blanc, Esq., in Suffolk, which are recorded by Professor Martyn. Martyn's Miller's Dict. art. Medicago.

47. MEDICAGO polymorpha. VARIABLE MEDIC.—This is also a plant much relished by cattle, but is not in cultivation: it is an annual, and perhaps inferior in many respects to the Nonsuch, which it in some measure resembles. There are many varieties of this plant cultivated in flower gardens on account of the curious shapes of the seed-pods, some having a distant resemblance to snails' horns, cater-pillars, &c. under which names they are sold in the seed-shops. It grows in sandy hilly soils; the wild kind has flat pods.

48. MEDICAGO sativa. LUCERNE.—Too much cannot be said in praise of this most useful perennial plant: it is every thing the farmer can wish for, excepting that it will not grow without proper culture. It should be drilled at eighteen inches distance, and kept constantly hoed all summer, have a large coat of manure in winter, and be dug into the ground between the drills. Six or seven pounds of seed will sow an acre in this mode.

I have known Lucerne sown with Grass and Clover for forming meadow land; but as it does not thrive well when encumbered with other plants, I see no good derived from this practice. No plant requires, or in fact deserves, better cultivation than this, and few plants yield less if badly managed.

49. MEDICAGO lupulina. TREFOIL, or NONSUCH.—A biennial plant, very usefully cultivated with Rye-grass and Clover for forming artificial meadows. Trefoil when left on the ground will seed, and these will readily grow and renew the plant successively; which has caused some persons to

suppose it to be perennial. About eight or ten pounds of seed are usually sown with six or eight pecks of Rye-grass for an acre, under a crop of Barley or Oats.

50. PLANTAGO lanceolata. RIB-GRASS.—This is a perennial plant, and very usefully grown, either mixed with grasses or sometimes alone: it will thrive in any soil, and particularly in rocky situations. It is much grown on the hills in Wales, where by its roots spreading from stone to stone it is often found to prevent the soil from being washed off, and has been known to keep a large district fertile which would otherwise be only bare rock. Sheep are particularly fond of it. About four pounds sown with other seeds for pasture, will render a benefit in any situation that wants it. Twenty-four pounds is usually sown on an acre when intended for the sole crop, and sown under corn.

51. POTERIUM Sanguisorba. BURNET.—This plant grows in calcareous soils, and is in some places much esteemed. On the thin chalky soils near Alresford in Hampshire, I have observed it to thrive better than almost any other plant that is cultivated. Sheep are particularly fond of it; and I have heard it said that the flavour of the celebrated Lansdown mutton arises from the quantity of Burnet growing there. It is also the favourite food of deer. This will grow well in any soil, and there are few pastures without it but would be benefited by its introduction. Twenty-five pounds per acre are sown alone: eight pounds mixed with other seeds would be sufficient to give a good plant on the ground.

52. SANGUISORBA officinalis. GREAT CANADA BURNET.—Cattle will eat this when young; and it has been supposed to be a useful plant, but I do not think it equal to Burnet.

It is perennial, and is often found wild, but has not yet been cultivated.

53. TRIFOLIUM pratense. RED CLOVER.—This is a very old plant in cultivation, and perhaps, with little exception, one of the most useful. It is very productive and nutritive, but soon exhausts the soil; and unless it is in particular places it presently is found to go off, which with the grazier is become a general complaint of all our cultivated Clovers. It is also well known, that if the crop is mown the plant is the sooner exhausted.

Seeds of Clover have the property of remaining long in the ground after it has become thus in a manner exhausted; and it frequently occurs that ashes being laid on will stimulate the land afresh, and cause the seeds to vegetate; which has given rise to the erroneous opinion with many persons, that ashes, and particularly soap ashes, will, when sown on land, produce Clover.

Red Clover is usually cultivated in stiff clays or loamy soils; and when sown alone, about sixteen or eighteen pounds of seed are used for the acre.

54. TRIFOLIUM medium. ZIGZAG, or MOUNTAIN-CLOVER.—Is in some degree like the preceeding; it produces a purple flower, and the foliage is much the same in appearance: but this is a much stronger perennial, and calculated from its creeping roots to last much longer in the land. It is equally useful as a food for cattle, and does not possess that dangerous quality of causing cattle to be hove, or blown, by eating it when fresh and green. This plant is, however, only to be met with in upland pastures, and there in its wild state; for it does not seed very abundantly, and is not in cultivation.

In the London seed-markets we often hear of a species of red Clover termed Cow-grass, and it generally sells for more money, and is said to differ in having the characters ascribed to it of this plant, namely, a hollow stem; the leaves more sharply pointed; the plant being a stronger perennial, and having the property of not causing the above-mentioned disorder to cows that eat of it. It is said to be cultivated in Hampshire, from whence I have often received the seeds which have been purchased purposely for the experiment; but on growing them, I never could discover these differences to exist. It is a circumstance worthy notice, that the very exact character of the Trifolium medium should thus be said to belong to the supposed variety of red Clover. I have endeavoured for the last twenty years to find out the true Cow-grass, and am of opinion that it has been from some cause mistaken for this plant.

The Trifolium medium is, at all events, a plant worth attention, and I think it might be easily brought into cultivation; for although it does not seed so abundantly as the T. pratense, I have observed it in places where a considerable quantity has been perfected, and where it might have been easily collected by gathering the capsules.

55. TRIFOLIUM repens. DUTCH CLOVER.—This is not so robust a plant as either of the former kinds, but it creeps on the ground and forms a fine bottom in all lands wherever it occurs, either cultivated or wild. This has not the property of blowing the cattle in so great a degree as the other sorts have. This disease is said to be accelerated by clover being eaten whilst the dew is on it: and when green clover is intended to be used as fodder, it is always best to mow it in the heat of the day, and let it lie till it is whithered, when it may be given to cows with safety.

Clover seeds of all kinds are necessary ingredients in laying down land to pasture; and the usual quantity is about twelve pounds per acre mixt in proportion at the option of the grower.

This kind remains longer in slight soils than the red does; but although both are perennial plants, they are apt to go off, for the reason pointed out under the head of T. pratense. This plant, as well as the T. medium and other perennial kinds, is sometimes found in old pastures on loamy soils; and whenever this is the case, it is a certain indication of the goodness of the soil, and such as a judicious gardener would make choice of for potting his exotic plants in, as he may rest assured that the soil which will maintain clover for a succession of seasons will be fit loam for such purposes.

56. TRIFOLIUM procumbens. YELLOW SUCKLING.— An annual very like the Nonsuch; it is a very useful plant, seeding very freely in pastures and growing readily, by which means it is every year renewed, and affords a fine bite for sheep and cattle. I have now and then seen the seeds of this in the shops, but it is not common. There is a gentleman who cultivates this plant very successfully near Horsham, and who, I am informed, states it to be the best kind of Clover for that land. It grows very commonly amongst the herbage on Horsham Common, so that it is probably its native habitat. The seeds are the smallest of all the cultivated Clovers, and of course less in weight will be necessary for the land.

57. TRIFOLIUM ochroleucum. YELLOW CLOVER.— This is not a common plant, but it deserves the attention of the grazier. I believe it is not in cultivation. In the garden it stands well, and is a large plant. The herbage appears to be

as good as that of any other kind of Clover, and it might, if introduced, be cultivated by similar means.

58. TRIFOLIUM agrarium. HOP TREFOIL.—This is also a good plant, but not in cultivation; it is eaten by cattle in its wild state, is a perennial, and certainly deserves a trial with such persons who may be inclined to make experiments with these plants.

Buffalo Clover is a kind similar to Trifolium agrarium and Trifolium repens, and appears to me to be a hybrid plant. This has been sometimes sent to this country from America, and is a larger plant than either. It has, however, as far as I have grown it, the same property of exhausting the soil as all the other species possess, and is soon found to go off: it is not in cultivation to any large extent.

59. VICIA Cracca. TUFTED VETCH.—Persons who have most noticed this plant have imagined it might be introduced into cultivation. It is hardy, durable, nutritious, and productive; but, like the Yellow Vetchling, the seeds do not readily vegetate; the only way to cultivate it, therefore, would be by planting out the roots; which might be done, as they are easily parted and are to be procured in great plenty in the places where it grows wild.

60. VICIA sativa. VETCHES, FETCH, or TARE.—A very useful and common plant, of which we have two varieties known to the farmer by the name of Spring and Winter Tares: they are both annuals. The spring variety is a more upright growing plant, and much tenderer than the other: it is usually sown in March and April, and affords in general fine summer fodder.

The Winter Tares are usually sown at the wheat seed-time, remain all winter, and are usually cut in the spring, generally six weeks before the spring crop comes in. The Winter Tares are now considered a crop worth attention by the farmers near London, who sow them, and sell the crop in small bundles in the spring at a very good price. Tares are usually sown broadcast, about three bushels and a half to the acre. Persons should be careful in procuring the true variety for the winter sowing; for I have frequently known a crop fail altogether by sowing the Spring Tares, which is a more tender variety, at that season. It should be noticed that the seeds of both varieties are so much alike that the kinds are not to be distinguished; but the plants are easily known as soon as they begin to grow and form stems; the Spring kind having a very upright habit, and the Winter Tares trail on the ground. It is usual for persons wanting seeds of such to procure a sample; and by growing them in a hothouse, or forcing frame, they may soon be able to ascertain the kinds. Ellis in his Husbandry says, that if ewes are fed on Tares, the lambs they produce will invariably have red flesh.

61. VICIA sylvatica. WOOD VETCH.—A perennial plant growing in the shade; it seems to have all the good properties in general with the other sorts of Tares; but it is not in cultivation.

62. VICIA sepium. BUSH VETCH.—Is also a species much eaten by cattle in its wild state, but has not yet been cultivated: it nevertheless would be an acquisition if it could be got to grow in quantity.

So much having been said of the different kinds of Tares, perhaps some persons may be inclined to think that it would be superfluous to have more in cultivation than one or two sorts. To this I would beg leave to reply, that they do not all grow exactly in the same situations wild; and if they were cultivated, some one of them might be found to suit in certain lands better than others; and perhaps we never shall see our agriculture at the height of improvement, till by some public-spirited measure all those things shall be grown for the purposes of fair comparative experiment—an institution much wanted in this country.

HINTS AS TO THE LAYING DOWN LAND TO PERMANENT PASTURE.

Having endeavoured to explain as nearly as possible the nature and uses of the plants which are likely to improve our meadows and pastures; I shall proceed to describe the best approved mode of sowing the land, on which depends, in a great measure, the future success of the husbandman's labour.

Under the head Lolium perenne I observed the practice of sowing clovers and that grass with a crop of barley or oats, which is intended as an intermediate crop for a season or two, and then the land to be again broken up and used for arable crops. And this is a common and useful practice; for although neither the Clover or Rye-grass will last long, yet both will be found to produce a good crop whilst the land will bear it, or until it is overpowered by the natural weeds of the ground [Footnote: It is not an uncommon opinion amongst farmers, that Rye-grass produces Couch; and this is not extraordinary; for, if the land is at all furnished with this weed, it receives great encouragement under this mode of culture.], which renders it necessary to the farmer to break it up.

I am aware of the difficulty of persuading persons (farmers in particular) to adopt any new systems; and I have often, when speaking of this subject amongst men of enlightened understandings, been told it would be next to madness, to sacrifice the benefit of a crop of oats or barley when the land is in fine tilth, and whilst we can grow grass seeds underneath it.

"To this I reply, that there is no land whatever, when left for a few months in a state of rest, but will produce naturally some kind of herbage, good and bad; and thus we find the industry of man excited, and the application of the hoe and the weeder continually among all our crops, this being essential to their welfare. I cannot help, therefore, observing how extremely absurd it is to endeavour to form clean and good pasturage under a crop hat gives as much protection to every noxious weed as to the young grass itself. Weeds are of two descriptions, and each requires a very different mode of extermination: thus, if annual, as the Charlock and Poppy, they will flower among the corn, and the seeds will ripen and drop before harvest, and be ready to vegetate as soon as the corn is removed; and if perennial, as Thistles, Docks, Couch-grass, and a long tribe of others in this way, well known to the farmer, they will be found to take such firm possession of the ground that they will not be got rid of without great trouble and expense.

"Although the crop of corn thus obtained is valuable, yet when a good and permanent meadow is wanted, and when all the strength of the land is required to nurture the young grass thus robbed and injured, the proprietor is often at considerable expense the second year for manure, which, taking into consideration the trouble and disadvantage attending it, more than counterbalances the profit of the corn crop.

"To accomplish fully the formation of permanent meadows, three things are necessary: namely to clean the land, to produce good and perfect seeds adapted to the nature of the soil, and to keep the crop clean by eradicating all the weeds, till the grasses have grown sufficiently to prevent the introduction of other plants. The first of these matters is known to every good farmer,—the second may be obtained,—and the third may be accomplished by practising the modes in which I have succeeded at a small comparative expense and trouble, and which is instanced in a meadow immediately fronting Brompton Crescent, the property of Angus Macdonald, Esq. which land was very greatly encumbered with noxious weeds of all kinds: but, by the following plan, the grasses were encouraged to grow up to the exclusion of all other plants; and though it has been laid down more than ten years, the pasturage is now at least equal to any in the county.

"Grass seeds may be sown with equal advantage both in spring and autumn. The land above mentioned was sown in the latter end of August, and the seed made use of was one bushel of Meadow-fescue, and one of Meadow fox-tail-grass, with a mixture of fifteen pounds of white Clover and Trefoil per acre; the land was previously cleaned as far as possible with the plough and harrows, and the seeds sown and covered in the usual way. In the month of October following, a most prodigious crop of annual weeds of many kinds having grown up, were in bloom, and covered the ground and the sown grasses; the whole was then mowed and carried off the land, and by this management all the annual weeds were at once destroyed, as they do not spring again if cut down when in bloom. Thus, whilst the stalks and roots of the annual weeds were decaying, the sown grasses were getting strength during the fine weather, and what few perennial weeds were amongst them were pulled up by hand in their young state. The whole land was repeatedly rolled, to prevent the worms and frost from throwing the plants out of the ground; and in the following spring it was grazed till the latter end of March, when it was left for hay, and has ever since continued a good field of grass.

"Several meadows at Roehampton, belonging to the late B. Goldsmid, Esq., were laid down with two bushels of Meadow fescue-grass and fifteen pounds of mixed Clover, and sown in the spring along with one peck and a half of

Barley, intended as a shade to the young grasses. The crop was thus suffered to grow till the latter end of June, and then the corn, with the weeds, was mowed and carried off the land; the ground was then rolled, and at the end of July the grasses were so much grown as to admit good grazing for sheep, which were kept thereon for several weeks. It should be observed, that the corn is to be mowed whilst in bloom, and when there is an appearance of, or immediately after rain; which will be an advantage to the grasses, and occasion them to thrive greatly.

"I sowed some fields for the same gentleman in autumn in the same way, and found them to succeed equally well."

The above remarks are part of a communication I gave six years since to the Society of Arts, for which I was honoured with their prize medal; and I have great pleasure in transcribing it [Footnote: See Transactions of the Society of Arts, vol. xxvii. p. 70.], as I frequently visit the meadows mentioned above, and have the satisfaction of hearing them pronounced the best in their respective neighbourhoods. Thus are my opinions on this head borne out by twelve years experience. Let the sceptic compare this improvement with his pretended advantage of a crop of Barley.

It should be observed that our agricultural efforts are intended only to assist the operations of nature, and that in all our experiments we should consult the soil as to its spontaneous produce, from whence alone we can be enabled to adapt, with propriety, plants to proper situations. The kinds of selected grass-seeds that are at this time to be purchased are few, and consist of Lolium perenne, Festuca pratensis, Alopecurus pratensis; Dactylis glomeratus, Cynosurus cristatus; with the various kinds of Clovers: and it is not easy to lay down any rule as to the mixture or proportion of each different kind that would best suit particular lands. Attention however should, in all cases, be paid to the plants growing wild in the neighbouring pastures, or in similar soils, and the greater portion used of those which are observed to thrive best.

In certain instances I have mentioned particular quantities of seeds to be mixed with others; but in general I have stated how much it would require to sow an acre with each kind separately; from which a person may form a criterion, when several sorts are used, as to what quantity of each sort should be adopted. Taking into view, therefore, that nothing but a mixture of proper kinds of Grasses, &c. will make good pasturage, and that our knowledge is very imperfect on this head at the present season, we must advise that particular attention be paid to the subject, or little good can be hoped for from all our endeavours.

SECT. III.—FODDER FROM LEAVES AND ROOTS.

The student in agriculture will find in this department a wide field for speculation, which, although it has been greatly improved during the last century, still affords much room for experiments.

During the last thirty-five years I have had opportunity of observing the great difference in the quantity of cattle brought to one of our largest beast-markets in the south of England; and it is well known that this has increased in a ratio of more than double; and I am informed by a worthy and truly honourable prelate, who has observed the same for twenty-five years previously, that it has nearly quadrupled. I have also made it my business, as a subject of curiosity, to inquire if the increase at other markets has been the same, and from all accounts I am convinced of the affirmative. Now as we have ample proofs from the statistical accounts of our husbandry, that less corn has not been grown in the same period, we shall naturally be inclined to give the merit of this increase to the introduction of the Turnip husbandry, which, although it is now become so general, is, comparatively speaking, but in its infancy; and it is from that branch of our agriculture that has sprung the culture of the great variety of fodder of the description which I am now about to explain.

And here it may not prove amiss to observe to the botanical student, should he hereafter be destined to travel, that by making himself thus acquainted with the nature of such vegetables, he may have it in his power to render great benefit to society by the introduction of others of still superior virtues, for the use both of man and the brute creation. When Sir Walter Raleigh undertook his expedition to South America, the object of which failed, he had the good fortune from his taste for botany to render to his country, and to the world at large, a more essential service, by the introduction of one single vegetable, than was ever achieved by the military exploits performed before or since that period [Footnote: The Potatoe was introduced by Sir Walter Raleigh, on his return from the River Plate, in the year 1586.]. It has not only been the means of increasing the wealth and strength of nations, but more than once prevented a famine in this country when suffering from a scarcity of bread-corn and when most of the ports which could afford us a supply were shut by the ambition of a powerful enemy.

63. BRASSICA Napus. TURNIP.—Turnips afford the best feed for sheep in the autumn and winter months. It is usual to sow them as a preparatory crop for Barley, and now very frequently for a crop of Spring Wheat. Turnips are not easily raised but where some kind of manure is used to stimulate the land. In dry seasons the crop is often destroyed by the ravages of a small beetle, which perforates the cotyledons of the plants, and destroys the crop on whole fields in a few hours.

Many remedies against this evil are enumerated in our books on husbandry. The best preventative, however, appears to be the putting manure on the ground in a moist state and sowing the seeds with it, in order to excite the young plant to grow rapidly; for the insect does not hurt it when the rough leaf is once grown. I have this season seen a fine field of Turnips, sown mixt with dung out of a cart and ploughed in ridges. The seeds which were not too deeply buried grew and escaped the fly; when scarcely a field in the same district escaped the ravages of that insect. Turnips are sown either broad-cast or in drills. It takes about four pounds of seed per acre in the first mode, and about half the quantity in the second.

There are several varieties of turnips grown for cattle; the most striking of which are, the White round Norfolk; the Red round ditto; the Green round ditto; the Tankard; the Yellow. These varieties are nearly the same in goodness and produce: the green and red are considered as rather more hardy than the others. The tankard is long-rooted and stands more out of the ground, and is objected to as being more liable to the attack of early frosts. The yellow is much esteemed in Scotland, and supposed to contain more nutriment [Footnote: The usual season for sowing the above varieties is within a fortnight or three weeks after Midsummer.]. The Stone and Dutch turnips are grown for culinary purposes, and are also sometimes sown after the corn is cleared, as being small and of early growth; these in such cases are called stubble turnips, and often in fine autumns produce a considerable quantity of herbage. For a further account of the culture &c. see Dickson's Modern Husbandry, vol. ii. p. 639.

There is nothing in husbandry requiring more care than the saving seeds of most of the plants of this tribe, and in particular of the Genus Brassica. If two sorts of turnips or cabbages are suffered to grow and bloom together, the pollen of each kind will be sufficiently mixed to impregnate each alternately, and a hybrid kind will be the produce, and in ninety-nine times out of a hundred a worse variety than either. Although this is generally the result of an indiscriminate mixture, yet by properly adapting two different kinds to grow together, new and superior varieties are sometimes produced. One gentleman having profited by this philosophy, has succeeded in producing some fine new varieties of fruits and vegetables, much to the honour of his own talents and his country's benefit [Footnote: See Mr

Knight On the Apple-tree.]. It is well known to gardeners that the cabbage tribe are liable to sport thus in their progeny; and to some accidental occurrence of this nature we are indebted for the very useful plant called the

64. ROOTA-BAGA. SWEDISH TURNIP.—Which is a hybrid plant par-taking of the turnip and cabbage, and what has within these few years added so much to the benefit of the grazier. This root is much more hardy than any of the turnips; it will stand our winters without suffering injury from frosts, and is particularly ponderous and nutritious.

It is usually cultivated as the common trunip, with this difference, that it requires to be sown as early in some lands as the month of May, it being a plant which requires a longer time to come to maturity.

Every judicious farmer who depends on turnips for foddering his stock in the winter, will do well to guard against the loss sometimes occasioned by the failure of his Turnips from frost and wet. Various ways of doing this are recommended, as stacking &c. But if he has a portion of his best land under Swedish turnip, he will have late in the winter a valuable crop that will be his best substitute. Another advantage is this, that it will last a fortnight longer in the spring, and consequently be valuable on this account. The quantity of seed usually sown is the same as for the common kinds of turnip. There are two varieties of this plant, one white and the other yellow: the latter is the most approved.

65. BRASSICA Napo Brassica. KOHLRABBI.—A hardy kind of Turnip cabbage, grown much in Germany for fodder: it is very nutritive, and has the property of resisting frost better than either the turnips or cattle-cabbage. The

seed and culture of this are the same as of Drum-head cabbage.

There are two varieties of this plant, the green and the purple; the latter is generally most esteemed.

66. BRUSSELS SPROUTS.—This is a large variety of cabbage, very productive and hardy. The culture is the same as for Cattle-cabbage.

67. BRASSICA oleracea. DRUM-HEAD CABBAGE.—This is usually sown in March and the plants put out into beds, and then transplanted into the fields; this grows to a most enormous size, and is very profitable. About four pounds of seed is sufficient for an acre.

SEC. IV.—GRAINS.

73. AVENA sativa. COMMON OATS.—A grain very commonly known, of which we have a number of varieties, from the thin old Black Oats to the fine Poland variety and the celebrated Potatoe-Oats.

These give the farmer at all times the advantage of a change of seeds, a measure allowed on all hands to be essential to good husbandry. The culture is various; thin soils growing the black kind in preference, which is remarkably hardy, where the finer sorts affecting a better soil will not succeed. It is applicable both to the drill and broad-cast. The seed is from six pecks to four bushels per acre, and the crop from seven to fourteen quarters.

74. CARUM Carui. CARAWAY SEEDS.—The seeds of this are in demand both by druggists and confectioners. It is cultivated in Kent and Essex; where it, being a biennial plant, is sown with a crop of spring corn, and left with the stubble during the succeeding winter, and after clearing the land in the spring is left to go to seed. It requires a good hot dry soil; but although the crop is often of great value, it so much exhausts the land as to be hazardous culture in many light soils where the dunghill is not handy.

The seed is about ten pounds per acre, and the crop often five or six sacks.

75. CORIANDRUM sativum. CORIANDER.—Is grown in the stiff lands, in Essex, and is an annual of easy but not of general culture. The seeds are used by druggists and rectifiers of spirits, and form many of the cordial drinks.

The quantity of seed and produce are similar to those of Caraway.

76. ERVUM Lens. LENTILS.—Once cultivated here for the seeds, which are used for soups; but it is furnished principally from Spain, and can at all times be purchased for less than it can be grown for.

77. HORDEUM distichon. COMMON TWO-ROWED BARLEY.—A grain now in very general cultivation, and supposed to be the best kind grown for malting. The season for sowing barley is in the spring, and the crop varies according to soil and culture; it is sown either broad-cast, drilled, or dibbled. The quantity of seed sown is from three pecks to three bushels per acre, and the produce from three to eleven quarters.

As the process of malting may not be generally understood by that class of readers for which this work is mostly intended, I shall give a short sketch of it.—It is a natural principle of vegetation, that every seed undergoes a change before it is formed into the young plant. The substance of the cotyledons, which when ground forms the nutritious flower of which bread is made, changes into two particular substances, i. e. sugar and mucilage; and whilst mankind form from it the principal staff of life as an edible commodity, the same parts of the seed in barley are by certain means made into malt, which is only another term for the sugar of that grain. To effect this, the barley is steeped in water, and afterwards laid in heaps, in which state it vegetates in a few days, and the saccharine fermentation is by that means carried on to a certain pitch, when it is put on a kiln to which a fire is applied, and it is by that means dried. It is then perfect malt, and fit for the purpose of brewing.

Pearl and Scotch Barley, used for soup and medicinal purposes, are made from the grain by being put into a mill, which merely grinds off the husk. The Pearl barley is mostly prepared in Holland, but the Scotch is made near Edinburgh in considerable quantities. A description of an improved Mill for this purpose is to be seen in the Edinburgh Encyclopaedia, p. 283.

78. HORDEUM vulgare. BERE, BIG, or WINTER BARLEY.—This is a coarser grain than the Two-rowed Barley, and hence it is not so well adapted to the purpose of malting. It is grown on cold thin soils, being much hardier than the former.

It is now often sown in October, and in the month of May or June following it is mown and taken off the land for green fodder. The plants will notwithstanding this produce in August a very abundant crop of grain. Hence this is a valuable mode of culture for the farmer.

The other varieties of Barley are,

79. HORDEUM hexastichon. SIX-ROWED BARLEY.— This is also a coarse grain; and although it was once in cultivation here, it has been altogether superseded by the Bere, which is a better kind.

80. HORDEUM zeocriton. BATTLEDORE BARLEY.— This is a fine grain, but very tender, and not now in cultivation in this country.

NAKED BARLEY. The two first species sometimes produce a variety which thrashes out of the husks similar to wheat: these are very heavy and fine grain, but they are not in cultivation: for what reason I know not.

81. PANICUM miliaceum. MILLET.—Millet is of two kinds, the brown and yellow. They are sometimes sown in this country for feeding poultry, and also for dressing; i. e. it is divested of the husk by being passed through a mill, when it is equal to rice for the use of the pastrycook. The seed used is from one to two bushels per acre. This is more commonly grown in Italy, and on the shores of the Mediterranean sea, from which large quantities are annually exported to the more northern countries.

82. PAPAVER somniferum. MAW-SEED.—The large white Opium Poppy is grown for seed for feeding birds, and also for pressing the oil, which is used by painters. The heads are also used by the apothecaries; which see under the head Medicinal Plants. About two pounds of seed to the acre.

83. PHALARIS canariensis. CANARY-SEED.—This is grown mostly in the Isle of Thanet, and sent to London &c. for feeding canary and other song-birds, and considered a very profitable crop to the farmer. It is sown in April, and the quantity of seed is about one bushel and a half per acre.

84. PISUM sativum. THE PEA [Footnote: At the request of Sir John Sinclair I made an experiment, from directions given by a French emigrant, of mixing Pease with urine in which had been steeped a considerable quantity of pigeon's dung. In the course of twenty-four hours they had swoln very much, when they were put into the ground. An equal quantity were steeped in water; and the same quantity also that had not been steeped, were sown in three adjoining spots of land. There was a difference in the coming up of the crops, of some days in each; but that with the above preparation took the lead, and was by far the best crop on the ground. This is an experiment worth attending to. It is usual to prepare wheat in a similar way, but no other grain

that I have ever heard of.].—The Gray Hog-pea used to be the only one considered sufficiently hardy for culture in the fields; but since the improvement in our agriculture we have all the finer varieties cultivated in large quantities. The seed used is about two bushels and a half per acre, and the produce varies from three to ten quarters.

The varieties of Peas are many, but the principal ones used in agriculture are the Early Charlton Pea; the Dwarf Marrow; the Prussian Blue. All these are dwarf kinds; and as the demand for this article in time of war is great for the navy and army, if the farmer's land will suit, and produce such as will boil, they will fetch a considerably greater price in proportion.

The varieties that are found to boil are either used whole, or split, which is done by steeping them in water till the cotyledons swell, after which they are dried on a kiln and passed through a mill; which just breaking the husk, the two cotyledons fall apart.

85. POLYGONUM Fagopyrum. BUCK-WHEAT.—This is usually sown in places where pheasants are bred, as the seed is the best food for those birds; it is also useful for poultry and hogs. I have eaten bread and cakes made of the flower, which are also very palatable. Two bushels are usually sown per acre. The season is May; and it is often sown on foul land in the summer, as it grows very thick on the land, and helps to clean it by smothering all the weeds. The crop does not stand on the ground more than ten or twelve weeks.

86. SECALE cereale. RYE.—This is often grown for a spring crop of green food, by sowing it early in the autumn, as it is very hardy and is not affected by frost. It grows fast in the spring months, and affords a very luxuriant crop of

green fodder. Tares and Rye are frequently sown mixed together for the same purpose, and the Tares find a support in the stalks of the Rye, by which means they produce a larger crop than they make by themselves. The grain is the next in estimation to Wheat, and is frequently used for making bread. The quantity sown per acre is the same as Wheat.

87. SINAPIS nigra. BLACK MUSTARD.—This is grown in Essex in great quantities for the seeds, which are sold to the manufacturers of flower of mustard, and is considered better flavoured, stronger, and capable of keeping better, than the white kind for such purpose. It is also in use for various medicinal preparations; which see. About two bushels of seed sown broad-cast are sufficient for an acre.

This plant affords another striking instance of the care of Providence in preserving the species of the vegetable kingdom, it being noticed in the Isle of Ely and other places, that wherever new ditches are thrown out, or the earth dug to any unusual depth, the seeds of Black Mustard immediately throw up a crop. In some places it has been proved to have lain thus embalmed for ages.

Flower of mustard, which is now become so common on our tables, and which is an article of very considerable trade, is but a new manufacture. A respectable seedsman who lived in Pall-Mall was the first who prepared it in this state for sale. The seeds of the white sort had been used to be bruised in a mortar and eaten sometimes as a condiment, but only in small quantities.

When used fresh it is weak, and has an unpleasant taste; but after standing a few hours the essential oil unites with the water which is used, and it then becomes considerably stronger, and the flavour is improved. It is prepared by drying the seeds on a kiln and grinding them to a powder. As this article is become of considerable importance from the demand, it has occasioned persons to speculate in its adulteration, which is now I believe often practised. Real flower of mustard will bear the addition of an equal quantity of salt without its appearing too much in the taste. In an old work, Hartman's treasure of Health, I find it to have been practised by a noble lady of that time to make mustard for keeping, with sherry wine with the addition of a little sugar, and sometimes a little vinegar. Query, Is this, with the substitution of a cheaper wine, the secret of what is called Patent Mustard?

88. TRITICUM aestivum. SPRING WHEAT.—Wheat is a grain well known in most countries in Europe. It has been in cultivation for many ages. This species was introduced some years ago from the Barbary coast, and has been found very beneficial for sowing in the spring, when it often produces a large crop. It takes a shorter time to come to maturity than the other sorts; and as it is a more profitable crop to the farmer on good soils than Barley, it is frequently sown after Turnips are over. This has, perhaps, been one of the best improvements in Grain husbandry that was ever introduced, as it gives the grower great advantages which he could not have under the common culture of Wheat at the usual seed-time. This is little different in appearance from the Common White Wheat. But there was a small variety of it with rounder grains sent to the Board of Agriculture from the Cape of Good Hope about the year 1801, of which I saved a small quantity of seeds which was distributed among the members; and I have lately seen a sample of it in the hands of a gentleman in Devonshire, who speaks very highly of it as producing a large crop in a short time, and that the flower was so much esteemed, that the millers gave him a higher price for it than the finest samples at market of the other kinds would sell for. I

believe this variety is very scarce. It is now twelve years since I grew it, from which what I saw, and all other in cultivation, if any there are, have sprung.

89. TRITICUM compositum. EGYPTIAN WHEAT.—This is a species with branched ears, and commonly having as many as three and four divisions. It is much cultivated in the eastern countries, but has not been found to answer so well in this country as the common cultivated species.

90. TRITICUM hybernum. COMMON WHEAT.—Of this grain we have a number of varieties, which are grown according to the fashion of countries, differing in the colour of the ear and also of the grain. The most esteemed sorts are the Hertfordshire White and the Essex Red Wheat, which are both much cultivated and equally esteemed. The season for growing these kinds is usually September and October. The drill, dibble, and broad-cast modes are all used, as the land and convenience of the farmer happen to suit, and the produce varies accordingly; as does also the quantity of seed sown. From two pecks to two bushels and a half are sown on an acre.

Wheat is liable to the ravages of many terrestrious insects which attack its roots; and also some very curious diseases. One of these has been very clearly elucidated by our munificent patron of science, Sir Joseph Banks, in the investigation of a parasitical plant which destroys the blood of the stalk and leaves, renders the grain thin, and in some cases quite destroys the crop, which has done that gentleman's penetration great credit [Footnote: Sir Joseph Banks On the Blight in Corn.]. An equally extraordinary disease is the Smut, which converts the farinaceous parts of the grain to a black powder resembling smut: a cirumstance too well known to many farmers. Those who wish to consult the remedies recommended against this, may refer to The Annals of Agriculture, and most other books on the subject. It is usual with farmers to mix the Wheat with stale urine or brine, and to dry it by sifting it with slaked lime, which has the effect of causing it to vegetate quickly, and to prevent the attacks of many insects when the seed is first put into the ground. This is considered as productive of great benefit to the crop; but it is also to be remarked, that it is almost the only grain that is ever prepared with this mixture, although it might be applied with equal propriety to all others. See article Pisum sativum.

91. TRITICUM turgidum. CONE WHEAT.—This a fine grain, and cultivated much in the strong land in the Vale of Evesham, where it is found to answer better than any other sorts. It is distinguished by the square and thick spike, and having a very long arista or beard.

The following sorts of Wheat are mentioned as being in cultivation. But I have not seen them, neither do I think any of them equal to the sorts enumerated above:

Triticum nigrum. BLACK-GRAINED WHEAT.
Triticum polonicum. POLISH WHEAT.
Triticum monococcon. ONE-GRAINED WHEAT.
Triticum Spelta. SPELT WHEAT.

Besides the use of Wheat for bread and other domestic purposes, large quantities are every season consumed in making starch, which is the pure fecula of the grain obtained by steeping it in water and beating it in coarse hempen bags, by which means the fecula is thus caused to exude and diffuse through the water. This, from being mixed with the saccharine matter of the grain, soon runs into the acetous fermentation, and the weak acid thus formed by digesting on the fecula renders it white. After setting, the precipitate is washed several times, and put by in square cakes and dried on kilns. These in drying part into flakes, which gives the form to the starch of the shops.

Starch is soluble in hot water, and becomes of the nature of gum. It is however insoluble in cold water, and on this account when pulverized it makes most excellent hair-powder.

92. Vicia Faba. THE BEAN.—Several kinds of Beans are cultivated by farmers. The principal are the Horse-Bean or Tick-Bean; the Early Mazagan; and the Long-pods. Beans grow best in stiff clayey soils, and in such they are the most convenient crop. The season for planting is either the winter or spring month, as the weather affords opportunity. They are either drilled, broad-cast sown, or put in by the dibble, which is considered not only the most eligible mode but in ge-neral affording the best crops. The seed is from one to three bushels per acre.

93. ZEA Mays. INDIAN CORN, or MAIZE. In warmer climates, as the South of France, and the East and West Indies, this is one of the most useful plants; the seeds forming good provender for poultry, hogs and cattle, and the green tops excellent fodder for cattle in general. I once saw a small early variety, that produced a very good crop, near Uxbridge; but I believe it is not in cultivation.

SECT. V.—MISCELLANEOUS ARTICLES.

94. CANNABIS sativa. HEMP.—This plant is cultivated in some parts of this country. It is usually sown in March, and is fit to harvest in October. It is then pulled up and immersed in water; when the woody parts of the stalks separating from the bark, which sloughs off and undergoes a decomposition by which the fibres are divided, it is then combed (hackled), dried, and reduced to different fineness of texture, and spun for various purposes. It requires good land, and the seed is usually two bushels and a half per acre.

The seed, which ripens about the time the hemp is pulled, is useful for feeding birds and poultry, and very nourishing.

95. DIPSACUS Fullonum. FULLER'S TEAVEL.—The heads of this plant are used for combing kerseymeres and finer broad cloths. The heads are generally fit to cut about the latter end of August, and are then separated and made up into bundles, and sold to the clothiers. The large heads are called Kings; the next size Middlings; and the smaller Minikins. The reason they are separated before sending to market is, that the large and small will not fit together on the frame in which they are fixed to the water-wheel, so that it is usual for the proprietor of the fulling-mills to purchase all of either one or the other size. The crop is considered very valuable, but the culture is confined to a small district in Somersetshire. The plant is biennial, and is usually sown in May, and the crop kept hoed during that season. In the following spring the plants bloom, and when the seeds are ripe the heads are fit for cutting; when they are assorted as above for the dealers. Three pounds of seed are used to an acre, and the plants at the last stirring are left from two feet to two feet and a half apart.

96. HUMULUS Lupulus. THE HOP.—The Hop is cultivated for brewing, being the most wholesome bitter we have, though the brewers are in the habit of using other vegetable bitters, which are brought from abroad and sold at a much cheaper rate. There is, however, a severe penalty on using any other than Hops for such purpose.

The Hops are distinguished by several varieties grown in Kent, Worcestershire, and at Farnham. The last place produces the best kind. For its culture more at length see Agriculture of Surry, by Mr. Stevenson.

97. ISATIS tinctoria. WOAD.—Is cultivated in the county of Somersetshire. It is used, after being prepared, for dyeing &c. It is said to be the mordant used for a fine blue on woollen. The foliage, which is like Spinach, is gathered during the summer months, and steeped in vats of water. After some time a green fecula is deposited in the bottom of the water, which is washed, and made into cakes and sold for use.

It is a perennial plant, and found wild in great abundance near Guildford, where great quantities might be gathered for use, and where a great deal of the seed could be collected. Its culture is very similar to that of the Teazle, with this difference, it requires the hoe at work constantly all the summer months.

The two plants Weld and Woad from the similarity of names are frequently confounded with each other, and some of the best agricultural writers have fallen into this error. They are two very different plants, and ought to be well defined, being each of them of very material consequence in this country.

98. LINUM usitatissimum. FLAX, or LINT-SEED.—Is grown for the purpose of making cloth, and has been considered a very profitable crop. The culture and management is similar to that of Hemp, and the seeds are in great demand for pressing. Lintseed oil, which it produces, is much used by painters, and is the only vegetable oil that is found fit for such purposes in general. The seeds are of several uses to the farmer; a tea is made of it, and mixed with skimmed milk, for fattening house-lambs and calves. Oxen are often fattened on the seed itself; but the cakes after the oil is expressed are a very common and most excellent article for fattening both black cattle and sheep. These are sold at from 10 l. to 16 l. per thousand.

It will require three bushels of Flax-seed for one acre, as it must be sown thick on the land. Lintseed cake has been used also for manure; and I have seen fine crops of Turnips where it has been powdered and sown in the drills with the seed.

99. RESEDA luteola. DYER'S-WEED, or WELD.—Is often confounded with Woad, but is altogether a very different plant. Weld is cultivated on the chalky hills of Surry, being sown under a crop of Barley, and the second year cleaned by hoeing, and then left to grow till it blooms, when it is pulled and tied up in small bundles, and after drying is sent to market, where it is purchased for dyeing yellow, and is in great request.

100. RUBIA tinctoria. MADDER.—This very useful dyeing drug used to be grown in this country in considerable quantities, but it is not cultivated here at the present time. The principal part of what is used now is brought from Holland, and affords a considerable article of trade to the Dutch farmers. Those who wish to be informed

of the mode of culture may consult Professor Martyn's edition of Miller's Dictionary.

Some years since Sir Henry Englefield, Bart., obtained a premium from the Society of Arts for the discovery of a fine tint drawn from Madder, called the Adrianople red. It was found that it was to be obtained from a variety of the Rubia brought from Smyrna; and Mr. Smyth, our consul at that city, was prevailed on by Dr. Charles Taylor to procure seeds from thence, which the Society did me the honour of committing to my care; and I have now a considerable stock of that kind, from whence I have myself obtained the same beautiful and superior tint. See Trans. Soc. Arts. vol. 27, p. 40.

101. ULEX europaeus. FURZE, GORSE, or WHIN.—Is used in husbandry for fences, and is also much cultivated for fuel for burning lime, heating ovens, &c. Cattle and sheep relish it much; but it cannot be eaten by them except when young, in consequence of its strong spines; to obviate which an implement has been invented for bruising it. When it grows wild on our waste land, it is common to set it on fire in the summer months, and the roots and stems will throw up from the ground young shoots, which are found very useful food for sheep and other animals. It is readily grown from seeds, six pounds of which will be enough for an acre of land.

PLANTS USEFUL IN THE ARTS.

SECT. VI.—BRITISH TREES AND SHRUBS.

102. ACER Pseudo-Platanus. SYCAMORE.—The wood of this tree is soft and of little use, unless it is for the turners' purposes, who make boxes and other small toys of it. It is not of value as timber.

103. ACER campestre. THE MAPLE.—Before the introduction of Mahogany and other fine woods the Maple was the principal wood used for all kinds of cabinet work, and was much esteemed: the knobs which grow on those trees in an old state afforded the most beautiful specimens, and according to Evelyn were collected by the curious at great prices. The Maple trees in this country are none of them at the present day old enough to afford that fine-veined variegation in the timber which is alluded to in this account.

104. ARBUTUS Unedo. THE STRAWBERRY-TREE.—Is a native of the islands in the celebrated Lake of Killarney in Ireland, where it grows to a large size. We know of no particular use to which it is applied. It is however one of our most ornamental evergreen shrubs, producing beautiful flowers, which vary from transparent white to deep red, in the winter months, at which season also the fruit appears; which taking twelve months to come to maturity affords the singular phaenomenon in plants, of having lively green leaves, beautiful flowers, and fruit as brilliant as the richest strawberry, in the very depth of our winter. We have a fine variety of this plant with scarlet blossoms, and also one

with double flowers, both of which are singularly ornamental to the shrubbery.

105. ARBUTUS Uva Ursi. BEAR-BERRIES.—A small trailing plant of great repute as a medicine, but of no use in any other respect.

106. BERBERIS vulgaris. BARBERRY.—This has long been cultivated in gardens for its fruit, which is a fine acid, and it is used as a conserve, and also for giving other sweeter fruits a flavour. The common wild kind has stones in the fruit, which renders it disagreeable to eat. There is a variety without stones called the Male Barberry, which is preferred on this account.

This tree is subject to a disease in the summer, caused apparently from a yellow fungus growing on the leaves and young shoots; and it is said that where it grows near corn fields it imparts its baneful influence to the grain, for which reason it is recommended in some of our books on agriculture to exterminate the trees.

107. BETULA alba. BIRCH-TREE.—Is in great use and of considerable value on some estates for making brooms, and the timber for all purposes of turnery-ware and carving. The sap of the Birch-tree is drawn by perforating the bark in the early state of vegetation. It is fermented, and makes a very pleasant and potent beverage called Birch Wine.

108. BETULA Alnus. ALDER-TREE.—This is a valuable tree for planting in moors and wet places. The wood is used for making clogs, pattens, and other such purposes; and the bark for dyeing and manufacturing some of the finer kinds of leather. This wood is of considerable value for making charcoal for gunpowder. In charring it a considerable

quantity of acetic acid is extracted, which is of great value for the purpose of bleaching, &c. &c.

109. BUXUS sempervirens. BOX-TREE.—The wood of Box is of great value for musical instruments, and for forming the handles of many tools: being very hard, it admits of a fine polish. This tree is growing in quantity at Box-hill in Surry, and has given name to that place.

This was planted by a late Duke of Norfolk, and has succeeded so well, that the wood has been cut twice, and sold each time for treble the value of the fee-simple of the land.

It forms a better cover for game than any other plant; and being very bitter, is not liable to be destroyed by any animal eating it down. An infusion of the leaves is frequently given as a vermifuge with good effect.

There is a smaller variety of this, much used for making edging to gravel walks in gardens.

110. CARPINUS Betulus. THE HORNBEAM.—This grows to a large tree, but is not of much account as timber: it is however very useful in forming ornamental fences, and is well adapted to this purpose from the tendency of its young branches to grow thick.

111. CLEMATIS Vitalba. TRAVELLER'S JOY.—A beautiful creeping shrub very useful to the farmers for making shackles for gates and hurdles, or withs for tying faggots and other articles. Whenever this plant is found in the hedges, &c. it is a certain indication of a ckalky under stratum in the soil.

112. CORNUS sanguinea. DOG-WOOD.—This is planted in pleasuregrounds as an ornamental shrub, and from the red appearance of the wood in the winter forms a beautiful constrast in plantations. It is also used by butchers for making skewers.

113. CORYLUS Avellana. THE HAZEL.—Is a well known shrub of large growth producing nuts, which are much admired. The Filbert is an improved variety of this plant. The farmers in Kent are the best managers of Filberts, and it is the only place where they are grown with any certainty; which appears to be owing principally to the trees being regularly pruned of the superfluous wood. It is performed in the month of March when the plants are in bloom, and is the only time when the fruit-bearing wood can be distinguished.

114. CRATAEGUS Aria. WHITE BEAM-TREE.—Is a beautiful tree producing very hard wood, and is much in esteem for cogs of millwork and various other purposes.

115. CRATAEGUS Oxyacantha. THE QUICKSET, or WHITE-THORN.—This is in great request for making fences, and is the best plant we know for such purposes if properly managed. It is readily propagated by sowing the hips, or fruit, which does not readily grow the first season; it is therefore usual to bury them mixed with saw-dust, or sand, one year, and then to sow them in beds.

116. DAPHNE Laureola. SPURGE- or WOOD-LAUREL.—Is used in medicine; which see.

We have many species of Daphne which are very ornamental to our shrubberies and green-houses: these are propagated principally by grafting; and the Wood-Laurel being hardy and of ready growth forms the stock principally used. It is readily propagated by seeds, which in three years will make plants large enough for this purpose.

The plant in all its parts is excessively acrid. I remember a man being persuaded to take the leaves reduced to powder, as a remedy for Syphilis, and he died in consequence in great agony in a few hours.

117. DAPHNE Mezerium. MEZERION.—Is a very beautiful shrub, and is one of the earliest productions of Flora, often exhibiting its brilliant scarlet flowers in January and February. We have also a white variety of this shrub in the gardens. The bark and roots are extremely acrimonious, and are used in medicine.

118. ERICA vulgaris. THE COMMON HEATH, HEATHER, or LING.—-This spontaneous produce of most of our sandy waste lands is of much usin rural oeconomy.

It is of considerable value for making brooms, and affords food to sheep, goats, and other animals; particularly to the grouse and heath-cock. The branches of heath placed upright in a wooden frame form the couch of repose to the brave Highlander. It is also stated that an excellent beverage was brewed from the tops of this plant, but the art of making it is now lost. This is the most common of the species, but all the others have similar properties. They are very ornamental plants. A numerous variety of heaths are brought from the Cape of Good Hope, and afford great pleasure to the amateur of exotic plants, being the greatest ornaments to our green-houses.

119. EUONYMUS europaeus. SPINDLE-TREE.—An ornamental shrub. The wood is in great request for making skewers for butchers, as it does not impart any unpleasant taste to the meat.

120. FAGUS Castanea. THE SPANISH CHESNUT.—This tree produces timber similar to oak in point of durability, and the bark also contains a considerable quantity of tannin. The Chesnut was in greater plenty in this country many years ago than at the present day; large forests are represented to have been in the neighbourhood of London; and we are led to believe such may have been the case, as many of the old buildings when examined have been found to be built of this timber. The fruit is used as a dainty at table; but the variety which is brought from Portugal and Spain is much larger than what are grown in this country. The large kind imported from those countries is grafted, and kept on purpose for the fruit. It is an improvement to graft this variety by taking the scions from trees in bearing, and they will produce fruit in a few years and in a dwarf state.

121. FAGUS sylvatica. THE BEECH.—The timber of the Beech is valuable for making wheels, and is applied to many other useful purposes in domestic oeconomy. The seeds of the Beech are very useful for fattening hogs.

This tree affords many beautiful varieties in foliage, the handsomest of which is the Copper Beech, whose purple leaves form a fine contrast in colour with the lively green of the common sort.

123. FRAXINUS excelsior. THE ASH.—The wood of the Ash is considered the best timber for all purposes of strong husbandry utensils. The wheels and axle-trees of carriages, the shafts for carts, and the cogs for mill-work, are principally made of this timber. The young wood when gown in coppices is useful for hop-poles, and the small underwood is said to afford the best fuel of any when used green. Coppice-land usually sells for a comparatively

greater price according as this wood prevails in quantity, on account of its good quality as fuel alone.

124. HEDERA Helix. IVY.—A common plant in woods, and often planted in shady places to hide walls and buildings. The leaves are good food for deer and sheep in winter. The Irish Ivy, which was brought from that country, is a fine variety with broad leaves. It was introduced by Earl Camden.

125. HIPPOPHAE Rhamnoides. SEA BUCKTHORN.—This is a scarce shrub; but is very useful as a plant for forming shelter on the hills near the sea-coast, it having been found to stand the sea-breeze better than any plant of the kind that is indigenous to this country.

126. ILEX aquifolium. HOLLY.—A well-known evergreen of singular beauty, of which we have many varieties, both striped, and of different colours in the leaf. Birdlime is made from the inner bark of this tree, by beating it in a running stream and leaving it to ferment in a close vessel. If iron be heated with charcoal made of holly with the bark on, the iron will be rendered brittle; but if the bark be taken off, this effect will not be produced. Ray's Works and Travels by Scott.

127. JUNIPERUS communis. JUNIPER.—An evergreen shrub, very common on waste lands. The berries are used in preparing the well-known spiritous liquor gin, and have been considered of great use in medicine.

128. LIGUSTRUM vulgare. PRIVET.—A shrub of somewhat humble growth, very useful for forming hedges where shelter is wanted more than strength. It bears clipping, and forms a very ornamental fence. There is a variety of this with berries, and another nearly evergreen.

129. MESPILUS germanica. THE MEDLAR.—Is cultivated for its fruit, and of which we have a variety called the Dutch Medlar; it is larger than our English one, but I do not think it better flavoured.

130. PINUS sylvestris. THE SCOTCH FIR.—A very useful tree in plantations for protecting other more tender sorts when young. It is also now very valuable as timber:—necessity, the common parent of invention, has taught our countrymen its value. When foreign deal was worth twenty pounds per load, they contrived to raise the price of this to about nine or ten pounds, and it was then thought proper for use; before which period, and when it could be bought for little money, it was deemed only fit for fuel. On the South Downs I know some plantations of this tree, which have been sold, after twenty-five years growth, at a price which averaged a profit of twenty shillings per annum per acre, on land usually let for sheep-pasture at one shilling and six-pence.

131. POPULUS alba. WHITE POPLAR. This is a very ornamental tree. The leaves on the under surface are of a fine white, and on the reverse of a very dark green; and when growing on large trees are truly beautiful, as every breath of air changes the colour as the leaves move. The wood of all the species of poplar is useful for boards, or any other purposes if kept dry. It is much in demand for floor-boards for rooms, it not readily taking fire; a red-hot poker falling on a board, would burn its way through it, without causing more combustion than the hole through which it passed.

132. POPULUS monilifera. CANADA POPLAR.—This is also known by the name of BLACK ITALIAN POPLAR, but from whence it had this name I do not know. This species, which is the finest of all the kinds, grows very

commonly in woods and hedges in many parts of Worcestershire and Herefordshire, where it reaches to prodigious sizes. Perhaps no timber is more useful than this; it is very durable, and easy to be converted to all purposes in building. The floors of a great part of Downton Castle, the seat of R. Payne Knight, Esq. are laid with this wood, which have been used forty years and are perfectly sound. Trees are now growing on his estate which are three and four feet in diameter. I have one growing in my Botanic garden which is eight years old, and measures upwards of six cubic feet of timber. The parent of this tree which grew at Brompton I converted into boards. It was nineteen years growing; and when cut down it was worth upwards of fourteen pounds, rating it at the then price of deal, for which it was a good substitute. Some fine specimens of this tree are also to be seen at Garnins, the seat of Sir J. G. Cotterell, Bart. the present worthy member for the county of Hereford.

133. PRUNUS domestica. THE COMMON PLUM-TREE.—This is the parent of our fruit of this name.

134. PRUNUS Cerasus. WILD CHERRY-TREE.—Is the parent of our fine cherries. It is cultivated much in Scotland for the timber, which is hard, and of use for furniture and other domestic purposes. It is the best and most lasting stock for grafting on. Persons who are about to plant this fruit would do well to inquire into the nature of the stock, as no fruit-tree is so liable to disease and become gummy as cherries are, and that is often much owing to the improved kinds being sown for stocks, which are of a more tender texture and of course less hardy than this.

135. PRUNUS insititia. SLOE-TREE.—Is of little use except when it occurs in fences. The fruit is a fine acid, and is much used by the common people, mixed with other

fruits less astringent and acid, to flavour made wines. It is believed that much Port wine is improved by the same means.

136. PYRUS communis. PEAR-TREE.—This is the parent of all our fine varieties of this fruit, and is used as the stock for propagating them; these are raised from seeds for that purpose. The wood of the Peartree is in great esteem for picture frames, it receiving a stain better than almost any other timber known.

137. PYRUS Malus. CRAB-TREE.—A tree of great account, as being the parent of all our varieties of apples, and is the stock on which the fine varieties are usually grafted. A dwarf variety of this tree, called the Paradise Apple, is used for stocks for making dwarf apple trees for gardens.

The juice of the Crab is called verjuice, which is in considerable demand for medicinal and other purposes.

138. QUERCUS robur. THE OAK.—Is a well known tree peculiar to Great Britain, and of the greatest interest to us as a nation. It is of very slow growth; but the timber is very strong and lasting, and hence it is used for building our shipping. The bark is supposed to contain more tannin than that of any other tree, and is valuable on that account. The acorns, or fruit, are good food for hogs, which are observed to grow very fat when turned into the forests at the season when they are ripe. The tree is raised from the acorn, which grows very readily.

We have accounts of Oak trees growing to great ages, and to most enormous sizes. One instance is mentioned by Evelyn, of one growing at Cowthorp, near Weatherby, in 1776, which within three feet of the ground was sixteen yards in circumference, and its height about eighty-five feet. Hunter's Evelyn's Sylva, p. 500.

139. ROSA rubiginosa. SWEET-BRIAR.—Is a very fragrant shrub, for which it has long been cultivated in the gardens. There are several varieties in the nurseries; as the Double-flowering, Evergreen, &c. which are much esteemed.

140. RUBUS Idaeus. THE RASPBERRY.—Produces a well known fruit in great esteem, and of considerable use both as food and for medicine.

141. RUBUS fruticosus. BRAMBLE.—Produces a black insipid fruit, but which is used by the poor people for tarts and to form a made wine: when mixt with the juice of sloes it is rendered very palatable.

142. RUBUS caesius.—Is a dwarf kind of bramble, and produces fruit of a pleasant acid, and where it grows in plenty it is used by the poor people for pies and other purposes of domestic oeconomy.

143. SALIX Russelliana. THE WILLOW.—No trees in this country are of more use than the species of this genus: many are grown for basket-makers in form of osiers, and other larger sorts serve for stakes, rails, hop-poles, and many other useful purposes. The bark of several species has been considered as useful for tanning leather. The charcoal of the Willow is also much in demand for making gunpowder.

144. SALIX viminalis. THE OSIER.—These are cultivated in watery places for making baskets, which are become a profitable article, and are the shoots of one season's growth

cut every winter. The species best adapted to this purpose, besides the common osier, are

The Salix vitellina. Golden Willow. The Salix monandria. Monandrous Willow. The Salix triandria. Triandrous Willow. The Salix mollissima. Silky-leaved Willow. The Salix stipularis. Auriculated Osier. The Salix purpurea. Bitter Purple Willow. The Salix Helix. Rose Willow. The Salix Lambertiana. Boyton Willow. The Salix Forbyana. Basket Osier. The Salix rubra. Green Osier. The Salix nigricans. Dark Purple Osier.

145. SAMBUCUS nigra. ELDER.—The timber of the Elder is useful for making musical instruments, and the berries made into wine and fermented make a useful and valuable beverage. A variety with green berries is much esteemed for wine also.

146. SORBUS Aucuparia. QUICKEN-TREE, or MOUNTAIN-ASH.—In this part of Britain we usually find this tree in plantations, where it is very ornamental; and the berries, which are of a fine scarlet, are the food of many species of birds. The wood is also useful for posts, &c. and is considered lasting.

147. SORBUS domestica. TRUE SERVICE.—Produces a fruit much like the Medlar, and when ripe is in great esteem. The only tree in this country in a wild state, is growing in Bewdley Forest, Worcester-shire.

148. SPARTIUM Scoparium. BROOM.—Is a very ornamental plant, and is used for making besoms. It was once considered as a specific in the cure of dropsy, but is now seldom used for medicial purposes.

149. STAPHYLEA pinnata. BLADDER-NUT.—This is not a common plant in this country. I know of no other use to which it is applied, but its being cultivated in nurseries and sold as an ornamental shrub. The seed-vessel, from whence it takes its name, is a curious example of the inflated capsule.

150. TAMARIX gallica. A shrub of large growth; and being less affected by the sea breeze than any others, is useful to form a shelter in situations where the bleak winds will not admit of trees of more tender kinds to flourish.

151. TAXUS baccata. THE YEW.—Was formerly much esteemed for making bows: but since those instruments of war and destruction have given place to the more powerful gun-powder, it is not so much in request. The wood is very hard and durable, and admits of a fine polish. The foliage of Yew is poisonous to cattle, who will readily eat it, if cut and thrown in their way in frosty weather.

152. TILIA europaea. THE LIME or LINDEN-TREE.—Is a very ornamental tree in plantations, and from its early putting forth its leaves is much esteemed. The flowers emit a very fine scent, and the inhabitants of Switzerland make a favourite beverage from them. The wood is very soft, though white and beautiful. It is much used for the ornamental boxes, &c. so well known by the name of Turnbridge-ware.

153. VACCINIUM uliginosum. GREAT BILBERRY. Vaccinium Vitis Idaea, RED WHORTLE-BERRY, and Vaccinium Oxycoccos, CRANBERRY, are all edible fruits, but do not grow in this part of the kingdom. Great quantities of Cranberries are imported every winter and spring from Russia; they are much esteemed by the confectioners for tarts, &c. and are sold at high prices.

These three kinds grow only in wet boggy places. A species which is native of America, called Vaccinium macrocarpon, has been very successfully cultivated at Spring Grove by Sir Joseph Banks, Bart. and which has also been attempted in various other places, but not with the same success. The fruit of this species is larger and of better flavour than either of the other kinds.

154. VACCINIUM Myrtillus. WHORTS, or BILBERRIES.—To a common observer this would appear to be a very insignificant shrub; it is not uncommonly met with on our heaths: but it is only in particular places where it fruits in abundance, and in such districts it is of considerable value.

The waste lands on Hindhead and Blackdown in Surry and Sussex are noticed for producing this fruit, which is similar to Black Currants. They are gathered in the months of August and September, and sold at the neighbouring markets.

In a calculation of the value of this plant with an intelligent nurseryman in that county, we found that from 500 l. to 700 l. were earned and realized annually by the neighbouring poor, who employed their families in this labour, and who are in the habit of travelling many miles for this purpose. The fruit is ripe in August, and at that season is met with in great plenty in all the neighbouring towns.

155. VISCUM album. MISSELTO.—A parasitical plant well known, and formerly of much repute in medicine, but wholly disregarded in the present practice. Birdlime is made from the berries.

Dr. Pulteney in tracing the history of Botanic science quotes Pliny for an account of the veneration in which this plant was held by the Druids, who attributed almost divine efficacy to it, and ordained the collecting it with rites and ceremonies not short of the religious strictness which was countenanced by the superstition of the age. It was cut with a golden knife, and when the moon was six days old gathered by the priest, who was clothed with white for the occasion, and the plant received on a white napkin, and two white bulls sacrificed. Thus consecrated, Misselto was held to be an antidote to poison, and prevented sterility. Query, Has not the custom of hanging up Misselto at merry-makings, and the ceremony so well known among our belles, some relation to above sacrifice?

156. ULEX europaeus. COMMON FURZE.—The culture of this shrub is given in the Agricultural Plants, being good for feeding cattle; its principal use however is for fuel, and it is frequently grown for such purposes. It is common on most of our waste lands. It also forms good fences, but should always be kept short and young, otherwise it becomes thin, especially in good land where it grows up and makes large bushes.

157. ULMUS campestris. THE ELM.—We have a number of varieties of the Elm; the most esteemed is that with the smooth bark. The timber has been long in request for water-pipes, and for boards, which are converted into various uses in domestic oeconomy.

158. ULMUS montana. BROAD-LEAVED ELM.—This has not been considered of so great value as the common sort, but it is of much more free growth; and I have been informed that in the West of England the timber has been found to be good and lasting.

SECT. VII.—PLANTS USEFUL IN MEDICINE.

The initial letters in this class distinguish the Pharmacopoeia in which each plant is inserted.

"By the wise and unchangeable laws of Nature established by a Being infinitely good and infinitely powerful,—not only man, the lord of the creation, 'fair form who wears sweet smiles, and looks erect on heaven,' but every subordinate being becomes subject to decay and death: pain and disease, the inheritance of mortality, usually accelerate his dissolution. To combat these, to alleviate when it has not the power to avert, Medicine, honoured art! comes to our assistance.

"It will not be expected that we should here give a history of this ancient practice, or draw a parallel betwixt the success of former physicians and those of modern times: all that concerns us to remark is, that the ancients were infinitely more indebted to the vegetable kingdom for the materials of their art than the moderns. Not so well acquainted with the oeconomy of nature, which teaches us that plants were chiefly destined for the food of various animals, they sought in every herb some latent healing virtue, and frequently endeavoured to make up the want of efficacy in one by the combination of numbers: hence the extreme length of their farraginous prescriptions. More enlightened ideas of the operations of medicine have taught the moderns greater simplicity and conciseness in practice. Perhaps there is a danger that this simplicity may be carried to far, and become finally detrimental to the practice."

The above is quoted from the Preface to a Catalogue of Medicinal Plants published by my predecessor in 1783: and it may be observed, that the medical student has, at the present season, a still less number of plants to store up in memory, owing, probably, to the great advances that chemistry has made in the mean time, through which mineral articles in many instances have superseded those of the vegetable kingdom. But, nevertheless, as Dr. Woodville has justly observed, "it would be difficult to show that this preference is supported by any conclusive reasoning drawn from a comparative superiority of the former;" or that the more general use of them has led to greater success in the practice of the healing art. It is however evident, that we have much to regret the almost total neglect of the study of medical botany by the younger branches of the professors of physic, when we are credibly informed that Cow-parsley has been administered for Hemlock, and Foxglove has been substituted for Coltsfoot [Footnote: See the account of a dreadful accident of this nature, in Gent. Mag. for Sept. 1815.], from which circumstance, some valuable lives have been sacrificed. It is therefore high time that those persons who are engaged in the business of pharmacy should be obliged to become so far acquainted with plants, as to be able to distinguish at sight all such as are useful in diet or medicine, and more particularly such as are of poisonous qualities.

The medical student has so many subjects for his consideration, that it is not desirable he should have a greater number of vegetables to consult than are necessary. And we cannot help lamenting the difficulty he has to struggle with in consequence of the great difference of names which the Pharmacopoeias of the present day exhibit. The London, Edinburgh, and Dublin, in many instances, enforce the necessity of learning a different term in each for the same thing, and none of which are called by the same they were twenty years ago. Surely it would be the means of forwarding the knowledge of drugs, if each could be distinguished by one general term.

The candidate for medical knowledge, however, is not the only one who has at times to regret this confusion of names. The Linnaean system is an easy and delightful path to the knowledge of plants; but, like all other human structures, it has its imperfections, and some of which have been modified by judicious alterations. Yet the teachers of this science, as well as the students, have often to deprecate the unnecessary change in names which has been made by many writers, though., in many cases, no more reason appears for it than there generally would be to change Christian and surnames of persons.

In the following section, I shall enumerate and describe those plants which are contained in the lists of the three colleges; and afterwards a separate list of those which, although they have been expunged, are still sometimes used by medical men.

I shall also endeavour to give such descriptions as are concise, at the same time sufficient for general knowledge, and for which reason I have taken Lewis's Materia Medica for my text, unless where improvements have been made in certain subjects I have consulted more modern authorities. It should be observed, that writers on medical plants, with few exceptions, have copied from one another: or with a little alteration as to words only.

And as some vegetables, from their affinitiy, may be confounded with others, whereby those possessing medical qualities may be substituted for others having none, or even poisonous ones, I shall in some instances enumerate a list of similar plants, which, with attention to their botanical characters, it is hoped will prevent those dangerous errors we have lately witnessed. As it is our business, in demonstrating plants, to guard the student against such confusion, it will be proper that specimens of such as come under this head be preserved, as a work for reference and contrast wherever doubts may arise.

158. ACONITUM Napellus. COMMON BLUE MONKSHOOD. The Leaves. L. E.—Every part of the fresh plant is strongly poisonous, but the root is unquestionably the most powerful, and when chewed at first imparts a slight sensation of acrimony, and a pungent heat of the lips, gums, palate and fauces, which is succeeded by a general tremor and sensation of chilliness.

This plant has been generally prepared as an extract or inspissated juice, after the manner directed in the Edinburgh and many of the foreign Pharmacopoeias, and, like all virulent medicines, it should be first administered in small doses. Stoerck recommends two grains of the extract to be rubbed into a powder with two drums of sugar, and as a dose to begin with ten grains of this powder two or three times a-day.

Similar Plants.—Aconitum japonicum; A. pyrenaicum; Delphinium elatum; D. exallatum.

Instead of the extract, a tincture has been made of the dried leaves macerated in six times their weight of spirit of wine, and forty drops given for a dose.—Woodville's Med. Bot. 965.

The Dublin College has ordered the Aconitum Neomontanum, which is not common in this country [Footnote: In plants of so very poisonous a nature as the Aconite, it is the duty of every one who describes them to be particular. Here seems to have been a confusion. The A.

Neomontanum is figured in Jacquin's Fl. Austriaca, fasc. 4. p. 381; and the first edition of Hortus Kewensis under A. Napellus erroneously quotes that figure: but both Gmelin in Syst. Vegetabilium, p. 838, and Wildenow in Spec. Plant. p. 1236, quote it under its proper name, A. Neomontanum. Now the fact is, that the Napellus is the Common Blue Monkshood; and the Neomontanum is altogether left out of the second edition of the Hortus Kewensis for the best of all reasons, it is not in this country; or, if it is, it must be very scarce, and, of course, not the plant used in medicine.].

160. ACORCUS Calamus. SWEET RUSH. The Root. L.— It is generally looked upon as a carminative and stomachic medicine, and as such is sometimes made use of in practice. It is said by some to be superior in aromatic flavour to any other vegetable that is produced in these northern climates; but such as I have had an opportunity of examining, fell short, in this respect, of several of our common plants. It is, nevertheless, a sufficiently elegant aromatic. It used to be an ingredient in the Mithridate and Theriaca of the London Pharmacopoeia, and in the Edinburgh. The fresh root candied after the manner directed in our Dispensatory for candying eryngo root, is said to be employed at Constantinople as a preservative against epidemic diseases. The leaves of this plant have a sweet fragrant smell, more agreeable, though weaker, than that of the roots.—Lewis's Mat. Med.

161. AESCULUS Hippocastanum. HORSE-CHESNUT. The Bark and Seed. E. D.— With a view to its errhine power, the Edinburgh College has introduced the seeds into the Materia Medica, as a small portion of the powder snuffed up the nostrils readily excites sneezing; even the infusion or decoction of this fruit produces this effect; it has therefore been recommended for the purpose of producing

a discharge from the nose, which, in some complaints of the head and eyes is found to be of considerable benefit.

On the continent, the Bark of the Horse Chesnut-tree is held in great estimation as a febrifuge; and, upon the credit of several respectable authors, appears to be a medicine of great efficacy.—Woodville's Med. Bot. 615.

162. AGRIMONIA Eupatoria. COMMON AGRIMONY. The Herb. D.—The leaves have an herbaceous, somewhat acrid, roughish taste, accompanied with an aromatic flavour. Agrimony is said to be aperient, detergent, and to strengthen the tone of the viscera: hence it is recommended in scorbutic disorders, in debility and laxity of the intestines, &c. Digested in whey, it affords an useful diet-drink for the spring season, not ungrateful to the palate or stomach.

163. ALLIUM Porrum. LEEK. The Root. L.—This participates of the virtues of garlic, from which it differs chiefly in being much weaker. See the article ALLIUM.

164. ALLIUM sativum. GARLIC. The Root. L. E. D.— This pungent root warms and stimulates the solids, and attenuates tenacious juices. Hence in cold leucophelgmatic habits it proves a powerful expectorant, diuretic, and emmenagogue; and, if the patient is kept warm, sudorific. In humoral asthmas, and catarrhous disorders of the breast, in some scurvies, flatulent colics, hysterical and other diseases proceeding from laxity of the solids, and cold sluggish indisposition of the fluids, it has generally good effects: it has likewise been found serviceable in some hydropic cases. Sydenham relates, that he has known the dropsy cured by the use of garlic alone; he recommends it chiefly as a warm strengthening medicine in the beginning of the disease.

Garlic made into an unguent with oils, &c. and applied externally, is said to resolve and discuss cold tumors, and has been by some greatly esteemed in cutaneous diseases. It has likewise sometimes been employed as a repellent. Sydenham assures us, that among all the substances which occasion a derivation or revulsion from the head, none operate more powerfully than garlic applied to the soles of the feet: hence he was led to make use of it in the confluent small-pox about the eighth day, after the face began to swell; the root cut in pieces, and tied in a linen cloth, was applied to the soles, and renewed once a day till all danger was over.

165. ALLIUM Cepa. ONION. The Root. D.—These roots are considered rather as articles of food than of medicine: they are supposed to afford little or no nourishment, and when eaten liberally they produce flatulencies, occasion thirst, headachs, and turbulent dreams: in cold phlegmatic habits, where viscid mucus abounds, they doubtless have their use; as by their stimulating quality they tend to excite appetite, attenuate thick juices, and promote their expulsion: by some they are strongly recommended in suppressions of urine and in dropsies. The chief medicinal use of onions in the present practice is in external applications, as a cataplasm for suppurating tumours, &c.

166. ALTHAEA officinalis. MARSH-MALLOW. The Leaves and Root. L.—This plant has the general virtues of an emollient medicine; and proves serviceable in a thin acrimonious state of the juices, and where the natural mucus of the intestines is abraded. It is chiefly recommended in sharp defluxions upon the lungs, hoarseness, dysenteries, and likewise in nephritic and calculous complaints; not, as some have supposed, that this medicine has any peculiar power of dissolving or expelling the calculus; but as, by lubricating and relaxing the vessels,

it procures a more free and easy passage. Althaea root is sometimes employed externally for softening and maturing hard tumours: chewed, it is said to give ease in difficult dentition of children.

The officinal preparations are:-Decoctio Althaeae officinalis, and Syrupus Althaeae.

Similar Plants.—Malva officinalis; M. rotundifolia; M. mauritanica; Lavatera arborscens.

This root gives name to an officinal syrup [L. E.] and ointment [L.] and is likewise an ingredient in the compound powder of gum tragacanth [L. E.] and the oil and plaster of mucilages [L.] though it does not appear to communicate any particular virtue to the two last, its mucilaginous matter not being dissoluble in oils.—Lewis's Mat. Med.

167. AMYGDALUS communis. SWEET and BITTER ALMONDS. L. E. D.—The oils obtained by expression from both sorts of almonds are in their sensible qualities the same. The general virtues of these oils are, to blunt acrimonious humours, and to soften and relax the solids: hence their use internally, in tickling coughs, heat of urine, pains and inflammations: and externally in tension and rigidity of particular parts.

168. ANCHUSA tinctoria. ALKANET-ROOT. E. D.— Alkanet-root has little or no smell: when recent, it has a bitterish astringent taste, but when dried scarcely any. As to its virtues, the present practice expects not any from it. Its chief use is for colouring oils, unguents, and plasters. As the colour is confined to the cortical part, the small roots are best, these having proportionally more bark than the large.

169. ANETHUM graveolens. DILL. The Seeds. L.—Their taste is moderately warm and pungent; their smell aromatic, but not of the most agreeable kind. These seeds are recommended as a carminative, in flatulent colics proceeding from a cold cause or a viscidity of the juices. The most efficacious preparations of them are, the distilled oil, and a tincture or extract made with rectified spirit. The oil and simple water distilled from them are kept in the shops.—Lewis.

170. ANETHUM Foeniculum. FENNEL. Seeds. E.—These are supposed to be stomachic and carminative; but this, and indeed all the other effects ascribed to them, as depending upon their stimulant and aromatic qualities, must be less considerable than those of Dill, Aniseed, or Caraway, though termed one of the four greater hot seeds.—Woodville's Med. Bot. p. 129.

171. ANGELICA Archangelica. GARDEN ANGELICA. The Root, Leaves, and Seeds. E.—All the parts of Angelica, especially the roots, have a fragrant aromatic smell, and a pleasant bitterish warm taste, glowing upon the lips and palate for a long time after they have been chewed. The flavour of the seeds and leaves is very perishable, particularly that of the latter, which, on being barely dried, lose greatest part of their taste and smell: the roots are more tenacious of their flavour, though even these lose part of it upon keeping. The fresh root, wounded early in the spring, yields and odorous yellow juice, which slowly exsiccated proves an elegant gummy resin, very rich in the virtues of the Angelica. On drying the root, this juice concretes into distinct moleculae, which, on cutting it longitudinally, appear distributed in little veins: in this state they are extracted by pure spirit, but not by watery liquors.

This resin is considered one of the most elegant aromatics of European growth, though little regarded in the present practice, and is rarely met with in prescription; neither does it enter any officinal composition.

172. ANTHEMIS nobilis. CHAMOMILE. The Flowers. L.E.D.—These have a strong not ungrateful, aromatic smell, but a very bitter nauseous taste. They are accounted carminative, aperient, emollient, and in some measure anodyne: and stand recommended in flatulent colics, for promoting the uterine purgations, in spasmodic affections, and the pains of women in child-bed: sometimes they have been employed in intermittent fevers, and the nephritis. These flowers are also frequently used externally in discutient and antiseptic fomentations, and in emollient glysters. The double-flowered variety is usually cultivated for medicine, but the wild kind with single flowers is preferable.

Similar Plants.—Anthemis arvensis; A. Cotula; Pyrethrum maritimum.

173. ANTHEMIS Pyrethrum. PELLITORY OF SPAIN. The Root. L.—The principal use of Pyrethrum in the present practice is as a masticatory, for promoting the salival flux, and evacuating viscid humours from the head and neighbouring parts: by this means it very generally relieves the tooth-ach, pains of the head, and lethargic complaints. If a piece of the root, the size of a pea, be placed against the tooth, it instantly causes the saliva to flow from the surrounding glands, and gives immediate relief in all cases of that malady.

174. APIUM Petroselium. COMMON PARSLEY. The Root. E.—Both the roots and seeds of Parsley are directed by the London College for medicinal use: the former have a

sweetish taste, accompanied with a slight warmth of flavour somewhat resembling that of a carrot; the latter are in taste warmer and more aromatic than any other part of the plant, and also manifest considerable bitterness.

These roots are said to be aperient and diuretic, and have been employed in apozems to relieve nephritic pains, and obstructions of urine.

Although Parsley is commonly used at table, it is remarkable that facts have been adducted to prove, that in some constitutions it occasions epilepsy, or at least aggravates the epileptic fit in those who are subject to this disease. It has been supposed also to produce inflammation in the eyes.—Woodville's Med. Bot. p. 43. A variety which produces larger roots, called Hamburgh Parsley, is commonly grown for medicinal uses.

175. ARBUTUS Uva Ursi. TRAILING ARBUTUS or BEAR-BERRY. The Leaves.—This first drew the attention of physicians as an useful remedy in calculous and nephritic affections; and in the years 1763 and 1764, by the concurrent testimonies of different authors, it acquired remarkable celebrity, not only for its efficacy in gravelly complaints, but in almost every other to which the urinary organs are liable, as ulcers of the kidneys and bladder, cystirrhoea, diabetes, &c. It may be employed either in powder or decoction; the former is most commonly preferred, and given in doses from a scruple to a dram two or three times a-day.— Woodville's Med. Botany.

176. ARNICA montana. MOUNTAIN ARNICA. The whole Plant. E. D.—The odour of the fresh plant is rather unpleasant, and the taste acrid, herbaceous, and astringent; and the powdered leaves act as a strong sternutatory.

This plant, according to Bergius, is an emetic, errhine, diuretic, diaphoretic, emmenagogue; and from its supposed power of attenuating the blood, it has been esteemed so peculiarly efficacious in obviating the bad consequences occasioned by falls and bruises, that it obtained the appellation of Panacea Lapsorum.—Woodville's Med. Bot. p. 43.

177. ARTEMISIA Absinthium. WORMWOOD, The Herb. L.—Wormwood is a strong bitter; and was formerly much used as such against weakness of the stomach, and the like, in medicated wines and ales. At present it is rarely employed in these intentions, on account of the ill relish and offensive smell which it is accompanied with. These it may be in part freed from by keeping, and totally by long coction, the bitter remaining entire. An extract made by boiling the leaves in a large quantity of water, and evaporating the liquor with a strong fire, proves a bitter sufficiently grateful, without any disgustful flavour.

178. ARTEMISIA Abrotanum. SOUTHERNWOOD. Leaves. D.—Southernwood has a strong, not very disagreeable smell; and a nauseous, pungent, bitter taste; which is totally extracted by rectified spirit, less perfectly by watery liquors. It is recommended as an anthelmintic; and in cold lencophlegmatic habits, as a stimulant, detergent, aperient, and sudorific. The present practice has almost entirely confined its use to external applications. The leaves are frequently employed in discutient and antiseptic fomentations; and have been recommended also in lotions and unguents for cutaneous eruptions, and the falling off of the hair.

179. ARTEMISIA maritima. SEA WORMWOOD. Tops. D.—In taste and smell, it is weaker and less unpleasant than the common worm-wood. The virutes of both are

supposed to be of the same kind, and to differ only in strength.

The tops used to enter three of our distilled waters, and give name to a conserve. They are an ingredient also in the common fomentation and green oil.

180. ARTEMISIA Santonica. ROMAN WORMWOOD. Seeds. E. D.—It is a native of the warmer countries, and at present difficultly procurable in this, though as hardy and as easily raised as any of the other sorts. Sea wormwood has long supplied its place in the markets, and been in general mistaken for it.

Roman wormwood is less ungrateful than either of the others: its smell is tolerably pleasant: the taste, though manifestly bitter, scarcely disagreeable. It appears to be the most eligible of the three as a stomachic; and is likewise recommended by some in dropsies.

181. ARUM maculatum. BITING ARUM. Fresh Root. L. E.—This root is a powerful stimulant and attenuant. It is reckoned a medicine of great efficacy in some cachectic and chlorotic cases; in weakness of the stomach occasioned by a load of viscid phlegm, and in such disorders in general as proceed from a cold sluggish indisposition of the solids and lentor of the fluids. I have experienced great benefit from it in rheumatic pains, particularly those of the fixed kind, and which were seated deep. In these cases I have given from ten grains to a scruple of the fresh root twice or thrice a day, made into a bolus or emulsion with unctuous and mucilaginous substances, which cover its pungency, and prevent its making any painful impression on the tongue. It generally excited a slight tingling sensation through the whole habit, and, when the patient was kept warm in bed, produced a copious sweat.

The only officinal preparation, in which this root was an ingredient, was a compound powder; in which form its virtues are very precarious. Some recommend a tincture of it drawn with wine; but neither wine, water, nor spirit, extract its virtues.—Lewis's Mat. Med.

182. ASARUM Europaeum, ASARABACCA. The Leaves. L. E. D.—Both the roots and leaves have a nauseous, bitter, acrimonious, hot taste; their smell is strong, and not very disagreeable. Given in substance from half a dram to a dram, they evacuate powerfully both upwards and downwards. It is said that tinctures made in spirituous menstrua possess both the emetic and cathartic virtues of the plant: that the extract obtained by inspissating these tinctures acts only by vomit, and with great mildness: that an infusion in water proves cathartic, rarely emetic: that aqueous decoctions made by long boiling, and the watery extract, have no purgative or emetic quality, but prove notable diaphoretics, diuretics, and emmenagogues.

Its principal use at present is as a sternutatory. The root of asarum is perhaps the strongest of all the vegetable errhines, white hellebore itself not excepted. Snuffed up the nose, in the quantity of a grain or two, it occasions a large evacuation of mucus, and raises a plentiful spitting. The leaves are considerably milder, and may be used to the quantity of three, four, or five grains. Geoffroy relates, that after snuffing up a dose of this errhine at night, he has frequently observed the discharge from the nose to continue for three days together; and that he has known a paralysis of the mouth and tongue cured by one dose. He recommends this medicine in stubborn disorders of the head, proceeding from viscid tenacious matter, in palsies, and in soporific distempers. The leaves are an ingredient in the pulvis sternutatoris of the shops.

183. ASPIDIUM Filix-Mas. Polypodium, Linn. MALE FERN. The Roots. L. E. D.—They are said to be aperient and anthelmintic. Simon Pauli tells us, that they have been the grand secret of some empirics against the broad kind of worms called taenia; and that the dose is one, two, or three drams of the powder. Two other kinds of Ferns used to be recommended; but this, being the strongest, has therefore been made choice of in preference, though the College of Edinburgh still retain them in their Catalogue of Simples.—Lewis's Mat. Med.

184. ASTRAGALUS Tragacanthus. GOATS-THORN. The Gum. L. E. D.—This gum is of a strong body, and does not perfectly dissolve in water. A dram will give to a pint of water the consistence of a syrup, which a whole ounce of gum Arabic is scarce sufficient to do. Hence its use for forming troches, and the like purposes, in preference to the other gums. It is used in an officinal powder, and is an ingredient in the compound powders of ceruss and amber.—Lewis's Mat. Med.

185. ATROPA Belladonna. DEADLY NIGHTSHADE. The Leaves, L. E. D.— Belladonna was first employed as an external application, in the form of fomentation, to scirrhus and cancer. It was afterwards administered internally in the same affections; and numerous cases, in which it had proved successful, were given on the authority of the German practitioners. It has been recommended, too, as a remedy in extensive ulceration, in paralysis, chronic rheumatism, epilepsy, mania, and hydrophobia, but with so little discrimination, that little reliance can be placed on the testimonies in its favour; and, in modern practice, it is little employed. It appears to have a peculiar action on the eye: hence it has been used in amaurosis; and from its power of causing dilatation of the pupil, when topically applied under the form of infusion, it has been used before

performing the operation for cataract. A practice which is hazardous, as the pupil, though much dilated by the application, instantly contracts when the instrument is introduced. When given internally, its dose is from one to three grains of the dried leaves, or one grain of the inspissated juice.—Murray's Mat. Med. p. 174.

I have had a cancer of the lip entirely cured by it: a scirrhosity in a woman's breast, of such kind as frequently proceeds to cancer, I have found entirely discussed by the use of it. A sore, a little below the eye, which had put on a cancerous appearance, was much mended by the internal use of the Belladonna; but the patient having learned somewhat of the poisonous nature of the medicine, refused to continue the use of it; upon which the sore grain spread, and was painful; but, upon a return to the use of the Belladonna, was again mended to a considerable degree; when the same fears again returning, the use of it was again laid aside, and with the same consequence, the sore becoming worse. Of these alternate states, connected with the alternate use of and abstinence from the Belladonna, there were several of these alterations which fell under my own observation [Footnote: See the Poisonous Plants, in a future page].—Cullen's Mat. Med. vol. ii. p. 270.

186. CARDAMINE pratensis. LADIES SMOCK. The Leaves. L. E. D.—Long ago it was employed as a diuretic; and, of late, it has been introduced in nervous diseases, as epilepsy, hysteria, choraea, asthma, &c. A dram or two of the powder is given twice or thrice a-day. It has little sensible operation.

187. CARUM Carui. CARAWAY. The Seeds. L. E. D.— These are in the number of the four greater hot seeds; and frequently employed as a stomachic and carminative in flatulent colics, and the like. Their officinal preparations

are an essential oil and a spiritous water; they were used as ingredients also in the compound juniper water, tincture of sena, stomachic tincture, oxymel of garlic, electuary of bayberries and of scammony, and the cummin-seed plaster.

188. CENTAUREA benedicta. BLESSED THISTLE. The Leaves. E. D.—The herb should be gathered when in flower, great care taken in drying it, and kept in a very dry airy place, to prevent its rotting or growing mouldy, which it is very apt to do. The leaves have a penetrating bitter taste, not very strong or very durable, accompanied with an ungrateful flavour, which they are in great measure freed from by keeping.

The virtues of this plant seem to be little known in the present practice. We have frequently experienced excellent effects from a light infusion of carduus in loss of appetite, where the stomach was injured by irregularities. A stronger infusion made in cold or warm water, if drunk freely, and the patient kept warm, occasions a plentiful sweat, and promotes all the secretions in general.

The seeds of this plant are also considerably bitter, and have been sometimes used for the same purposes as the leaves.

189. CHIRONIA Centaurium. LESSER CENTAURY. The Tops. L. E. D.—This is justly esteemed to be the most efficacious bitter of all the medicinal plants indigenous to this country. It has been recommended as a substitute for Gentian, and, by several, thought to be a more useful medicine: experiments have also shown it to possess an equal degree of antiseptic power.

Many authors have observed, that, along with the tonic and stomachic qualities of a bitter, Centaury frequently proves cathartic; but it is possible that this seldom happens, unless it be taken in very large doses. The use of this, as well as of the other bitters, was formerly common in febrile disorders previous to the knowledge of Peruvian-bark, which now supersedes them perhaps too generally; for many cases of fever occur which are found to be aggravated by the Cinchona, yet afterwards readily yield to the simple bitters.—Woodville, p. 277.

190. COCHLEARIA officinalis. SCURVY-GRASS. The Herb. E.—Is antiseptic, attenuant, aperient, and diuretic, and is said to open obstructions of the viscera and remoter glands, without heating or irritating the system. It has long been considered as the most effectual of all the antiscorbutic plants; and its sensible qualities are sufficiently powerful to confirm this opinion. In the rheumatismus vagus, called by Sydenham Rheumatismus scorbuticus, consisting of wandering pains of long continuance, accompanied with fever, this plant, combined with Arum and Wood-Sorrel, is highly commended both by Sydenham and Lewis.

We have testimony of its great use in scurvy, not only from physicians, but navigators; as Anson, Linschoten, Maartens, Egede, and others. And it has been justly noticed, that this plant grows plentifully in those high latitudes where the scurvy is most obnoxious. Forster found it in great abundance in the islands of the South Seas.—Woodville, p. 395.

191. COCHLEARIA Armoracia. HORSE-RADISH. The Root. E.-The medical effects of this root are, to stimulate the solids, attenuate the juices, and promote the fluid secretions: it seems to extend its action through the whole habit, and affect the minutest glands. It has frequently done great service in some kinds of scurvies and other chronic

disorders proceeding from a viscidity of the juices, or obstructions of the excretory ducts. Sydenham recommends it likewise in dropsies, particularly those which sometimes follow intermittent fevers. Both water and rectified spirit extract the virtues of this root by infusion, and elevate them in distillation: along with the aqueous fluid an essential oil arises, possessing the whole taste and pungency of the horse-radish. The College have given us a very elegant compound water, which takes its name from this root.

192. COLCHICUM autumnale. MEADOW-SAFFRON. The Roots. L. E. D.—The roots, freed from the outer blackish coat and fibres below, are white, and full of a white juice. In drying they become wrinkled and dark coloured. Applied to the skin, it shows some signs of acrimony; and taken internally, it is said sometimes to excite a sense of burning heat, bloody stools, and other violent symptoms. In the form of syrup, however, it has been given to the extent of two ounces a-day without any bad consequence. It is sometimes employed as a diuretic in dropsy. It is now supposed to be a principal ingredient in the celebrated French gout medicine L'Eau Medicinale.

193. CONIUM maculatum. HEMLOCK. The Leaves. L. E. D.—Physicians seem somewhat in dispute about the best mode of exhibiting this medicine; some recommending the extract, as being most easily taken in the form of pills; others the powder, as not being subject to that variation which the extract is liable to, from being made in different ways. With respect to the period, likewise, at which the plant should be gathered, they seem not perfectly agreed; some recommending it when in its full vigour, and just coming into bloom, and others, when the flowers are going off. An extract of the green plant is ordered by the College in their last list. Dr. Cullen has for many years commended the making it from the unripe seeds; and this mode the

College of Physicians at Edinburgh have thought proper to adopt in their late Pharmacopoeia.

Similar Plants.—Aethusa Cynapium; Apium Petroselium; Oenanthe crocata; Oe. fistulosa; Phellandrium aquaticum.

194. CORIANDRUM sativum. CORIANDER. The Seeds. L. E. D.-These, when fresh, have a strong disagreeable smell, which improves by drying, and becomes sufficiently grateful. They are recommmended as carminative and stomachic.

195. CROCUS sativus. TRUE SAFFRON. The Stigmata. L. E. D.—There are three sorts of saffron met with in the shops, two of which are brought from abroad, the other is the produce of our own country. This last is greatly superior to the two former.

This medicine is particularly serviceable in hysteric depressions proceeding from a cold cause, or obstruction of the uterine secretions, where other aromatics, even those of the more generous kind, have little effect. Saffron imparts the whole of its virtue and colour to rectified spirit, proof spirit, wine, vinegar, and water: a tincture used to be drawn with vinegar, but it looses greatly its colour in keeping. There can be little use for preparations of saffron, as the drug itself will keep good for any length of time.

196. CUMINUM Cymini. CUMMIN. The Seeds. L.— Cummin seeds have a bitterish warm taste, accompanied with an aromatic flavour, not of the most agreeable kind. They are accounted good carminatives, but not very often made use of. An essential oil of them used to be kept in the shops, and they gave name to a plaster and cataplasm.— Lewis's Mat. Med.

197. CYNARA Scolymus. ARTICHOKE. The Leaves. E.—The bitter juice of the leaf, mixed with an equal part of Madeira wine, is recommended in an ounce dose night and morning, as a powerful diuretic in dropsy. An infusion of the leaf may likewise be used.

198. DAPHNE Mezereum. THE MEZEREON. The Roots. L. E. D.—This plant is extremely acrid, especially when fresh, and, if retained in the mouth, excites great and long continued heat and inflammation, particularly of the throat and fauces. The bark and berries of Mezereon in different forms have been long externally used to obstinate ulcers and ill conditioned sores. In France, the former is strongly recommended as an application to the skin, which, under certain management, produces a continued serious discharge without blistering, and is thus rendered useful in many chronic diseases of a local nature answering the purpose of what has been called a perpetual blister, while it occasions less pain and inconvenience.

In this country Mezereon is principally employed for the cure of some siphylitic complaints; and in this way Dr. Donald Monro was the first who gave testimony of its efficacy in the successful use of the Lisbon Diet Drink.

The considerable and long-continued heat and irritation that is produced in the throat when Mezereon is chewed, induced Dr. Withering to think of giving it in a case of difficulty of swallowing, seemingly occasioned by a paralytic affection. The patient was directed to chew a thin slice of the root as often as she could bear it, and in about a month recovered her power of swallowing. This woman had suffered the complaint three years, and was greatly reduced, being totally unable to swallow solids, and liquids but very imperfectly.—Woodville's Med. Bot. p. 720.

199. DATURA Stramonium. THORN APPLE. The whole Plant. E.—Dr. Woodville informs us, that an extract of this plant has been the preparation usually employed, and from one to ten grains and upwards a-day: but the powdered leaves after the manner of those directed for hemlock would seem, for the reason given, to be a preparation more certain and convenient.

It has been much celebrated as a medicine in epilepsy and convulsions and mania; but it is of a violent narcotic quality, and extremely dangerous in its effects.

Stramonium has been recommended, as being of considerable use in cases of asthma, on the authority of some eminent physicians of the East Indies; and the late Dr. Roxburgh has stated to me many instances wherein it had performed wonders in that dreadful malady.

The Datura Metal, Purple-flowered Thorn-apple, is much like the Stramonium, except in the flowers and the stalks being of a purple colour. I have made particular inquiry of Dr. Roxburgh if any particular kind was used in preference, and he said not; that both the above sorts were used; and, in fact, not only these, but the Datura Tatula, another species which grows wild there, and is cultivated in our stoves for the sake of its beautiful flowers, is also used for the same purposes.

The mode of using it was by cutting the whole plant up after drying, and smoking it in a common tobacco-pipe; and which, in some cases in this country also, has given great ease in severe attacks; and I know several persons who use it with good effect to this day. In vegetables of such powerful effects as this is known to have, great care ought to be taken in their preparation, which, I fear, is not always so much attended to as the nature of this subject requires

[Footnote: See Observations on and Directions for preparing and preserving Herbs in general, et the end of this section.].

200. DAUCUS sylvestris. WILD CARROT. The Seeds. L.—These seeds possess, though not in a very considerable degree, the aromatic qualities common to those of the umbelliferous plants, and hence have long been deemed carminative and emmenagogue; but they are chiefly esteemed for their diuretic powers, and for their utility in calculus and nephritic complaints, in which an infusion of three spoonfuls of the seeds in a pint of boiling water has been recommended; or the seeds may be fermented in malt liquor, which receives from them an agreeable flavour resembling that of the lemon-peel.—Woodville's Med. Bot. p. 132.

Similar Plants.—Sison Amonum; Daucus Carota.

201. DAUCUS Carota. CULTIVATED CARROT. The Roots. L. E. D.—The expressed juice, or a decoction of these roots, has been recommended in calculous complaints, and as a gargle for infants in aphtous affections or excoriations of the mouth; and a poultice of scraped carrots has been found an useful application to phagedenic ulcers, and to cancerous and putrid sores.

202. DELPHINIUM Staphis Agria. STAVES AGRIA. The Seeds. L. D.— Stavesacre was employed by the ancients as a cathartic, but it operates with so much violence both upwards and downwards, that its internal use has been, among the generality of practitioners, for some time laid aside. It is chiefly employed in external applications for some kinds of cutaneous eruptions; and for destroying lice and other insects; insomuch that it has from this virtue

received its name in different languages, Herba pedicularis, Herbe aux poux, Lauskraut, Lousewort.

203. DIANTHUS caryophyllus. CLOVE-PINK. The Petals. E.—These flowers are said to be cardiac and alexipharmac. Simon Paulli relates, that he has cured many malignant fevers by the use of a de-coction of them; which he says powerfully promoted sweat and urine without greatly irritating nature, and also raised the spirits and quenched thirst. The flowers are chiefly valued for their pleasant flavour, which is entirely lost even by light coction. Lewis says, the College directed the syrup, which is the only officinal preparation of them, to be made by infusion.

204. DIGITALIS purpurea. FOXGLOVE. The Leaves. L. E. D.—The leaves of Foxglove have a nauseous taste, but no remarkable smell. They have been long used externally to sores and scrophulous tumours with considerable advantage. Its diuretic effects, for which it is now so deservedly received into the Materia Medica, were entirely overlooked. To this discovery Dr. Withering has an undoubted claim; and the numerous cures of dropsy related by him and other practitioners of established reputation, afford incontestable proofs of its diuretic powers, and of its practical importance in the cure of those diseases. The dose of dried leaves in powder is from one grain to three twice a-day; but if a liquid medicine be preferred, a dram of the dried leaves is to be infused for four hours in half a pint of boiling water, adding to the strained liquor an ounce of any spiritous water. One ounce of this infusion given twice a-day is a medium dose; it is to be continued in these doses till it either acts upon the kidneys; the stomach, or the pulse, (which it has a remarkable power of lowering,) or the bowels.— Woodville's Med. Bot. p. 221.

This is now become a very popular medicine, but if used incautiously is attended with danger. Medical practitioners should make themselves perfectly acquainted with this plant, as the leaves are the only part used; and their not being readily discriminated when separated from the flowers, several accidents have occurred. In the Gent. Mag. for September 1815 is recorded a very extraordinary mistake, where the life of a child was sacrificed to the ignorance of a person who administered this instead of Coltsfoot; a plant so very dissimilar, that, had it not been well authenticated, I should not have believed the fact.

Similar Plants.—Verbascum nigrum; V. Thapsus; Cynoglossum officinale, or, after the above mistake, any other plant with a lanceolate leaf, we fear, may be confounded with it.

205. ERYNGIUM maritimum. SEA-HOLLY. Roots. D.—The roots are slender, and very long; of a pleasant sweetish taste, which on chewing for some time is followed by a light degree of aromatic warmth and acrimony. They are accounted aperient and diuretic, and have also been celebrated as aphrodisiac: their virtues, however, are too weak to admit them under the head of medicines. The candied root is ordered to be kept in the shops.—Lewis's Mat. Med.

206. FERULA assafoetida. ASSAFOETIDA. Gum. L. E. D.—This drug has a strong fetid smell, somewhat like that of garlick; and a bitter, acrid, biting taste. It looses with age of its smell and strength, a circumstance to be particularly regarded in its exhibition. It consists of about one-third part pure resin, and two-thirds of gummy matter; the former soluble in rectified spirit, the other in water. Proof-spirit dissolves almost the whole into a turbid liquor; the tincture in rectified spirit is transparent.

Assafoetida is the strongest of the fetid gums, and of frequent use in hysteric and different kinds of nervous complaints. It is likewise of considerable efficacy in flatulent colics; and for promoting all the fluid secretions in either sex. The ancients attributed to this medicine many other virtues which are at present not expected from it.—Lewis's Mat. Med.

207. FICUS Carica. COMMON FIG. Fruit. L. D.—The recent fruit completely ripe is soft, succulent, and easily digested, unless eaten in immoderate quantities, when it is apt to occasion flatulency, pain of the bowels, and diarrhoea. The dried fruit is pleasanter to the taste, and is more wholesome and nutritive. Figs are supposed to be more nutritious by having their sugar united with a large portion of mucilaginous matter, which, from being thought to be of an oily nature, has been long esteemed an useful demulcent and pectoral; and it is chiefly with a view of these effects that they have been medicinally employed.

208. FRAXINUS Ornus. MANNA. L. E. D.—There are several sorts of Manna in the shops. The larger pieces, called Flake Manna, are usually preferred; though the smaller grains are equally as good, provided they are white, or of a pale yellow colour, very light, of a sweet not unpleasant taste, and free from any visible impurities.

Manna is a mild agreeable laxative, and may be given with saftey to children and pregnant women: nevertheless, in some particular constitutions it acts very unkindly, producing flatulencies and distension of the viscera.—Lewis's Mat. Med.

209. GENTIANA lutea. YELLOW GENTIAN. Root. L. D.—This root is a strong bitter, and, as such, very frequently made use of in practice: in taste it is less

exceptionable than most of the other substances of this class: infusions of it, flavoured with orange peel, are sufficiently grateful. It is the capital ingredient in the bitter wine; and a tincture and infusion of it are kept in the shops.

Lewis mentions a poisonous root being mixed among some of the Gentian brought to London; the use of which occasioned in some instances death. This was internally of a white colour, and void of bitterness. There is no doubt but this was the root of the Veratrum album, a poisonous plant so similar, that it might readily be mistaken for it.—Lewis's Mat. Med.

210. GEUM urbanum. COMMON AVENS. Root. D.— This has a warm, bitterish, astringent taste, and a pleasant smell, somewhat of the clove kind, especially in the spring, and when produced in dry warm soils. Parkinson observes, that such as is the growth of moist soils has nothing of this flavour. This root has been employed as a stomachic, and for strengthening the tone of the viscera in general: it is still in some esteem in foreign countries, though not taken notice of among us. It yields, on distillation, an elegant odoriferous essential oil, which concretes into a flaky form.—Lewis's Mat. Med.

Similar Plants.—Geum rivale; G. intermedium.

211. GLYCYRRHIZA glabra. LIQUORICE. Root. L. D.— This is produced plentifully in all the countries of Europe: that which is the growth of our own is preferable to such as comes from abroad; this last being generally mouldy, which this root is very apt to become, unless kept in a dry place.

The powder of liquorice usually sold is often mingled with flower, and, I fear, too often with substances not quite so wholesome. The best sort is of a brownish yellow colour (the fine pale yellow being generally sophisticated) and of a very rich sweet taste, much more agreeable than that of the fresh root. Liquorice is almost the only sweet that quenches thirst.

This root is a very useful pectoral, and excellently softens acrimonious humours, at the same time that it proves gently detergent: and this account is warranted by experience. It is an ingredient in the pectoral syrup, pectoral troches, the compound lime waters, decoction of the woods, compound powder of gum tragacanth, lenitive electuary, and theriaca. An extract is directed to be made from it in the shops; but this preparation is brought chiefly from abroad, though the foreign extract is not equal to such as is made with proper care among ourselves.—Lewis's Mat. Med.

212. GRATIOLA officinalis. HEDGE-HYSSOP. Herb. E. D.—The leaves have a very bitter disagreeable taste: an infusion of a handful of them when fresh, or a dram when dried, is said to operate strongly as a cathartic. Kramer reports that he has found the root of this plant a medicine similar in virtue to Ipecacuanha.

Similar Plants.—Lythrum Salicaria; Scutellaria galericulata.

213. HELLEBORUS niger. BLACK HELLEBORE. Root. L.—The tase of Hellebore is acrid and bitter. Its acrimony, as Dr. Grew observes, is first felt on the tip of the tongue, and then spreads immediately to the middle, without being much perceived on the intermediate part: on chewing it for a few minutes, the tongue seems benumbed, and affected with a kind of paralytic stupor, as when burnt by eating any thing too hot.

Our Hellebore is at present looked upon principally as an alterative, and in this light is frequently employed, in small doses, for attenuating viscid humours, promoting the uterine and urinary discharges, and opening inveterate obstructions of the remoter glands: it often proves a very powerful emmenagogue in plethoric habits, where steel is ineffectual or improper. An extract made from this root with water, is one of the mildest, and for the purposes of a cathartic the most effectual preparation of it: this operates sufficiently, without occasioning the irritation which the pure resin is accompanied with. A tincture drawn with proof-spirit contains the whole virtue of the Hellebore, and seems to be one of the best preparations of it: this tincture, and the extract, used to be kept in the shops. The College of Edinburgh used to make this root an ingredient in the purging cephalic tincture, and compound tincture of jalap; and its extract, in the purging deobstruent pills, gamboge pills, the laxative mercurial pills, and the compound cathartic extract.—Lewis's Mat. Med.

Similar Plant.—Helleborus viridis.

214. HELLEBORUS foetidus. BEARSFOOT. Leaves. L.— The root is a strong cathartic; it destroys worms, and is recommended in different species of mania. It is commonly substituted for that of the Helleborus viridis, which is a more dangerous medicine. Hill's Herbal, p. 32. Great care ought to be used in the administering this plant: many instances of its dreadful effects are related. (See Poisonous Plants.)

Similar Plant.—Helleborus viridis.

215. HORDEUM distichon. PEARL BARLEY. Seeds. L. E.—Barley, in its several states, is more cooling, less glutionous, and less nutritious than wheat or oats; among

the ancients, decoctions of it were the principal aliment, and medicine, in acute diseases. The London College direct a decoction of pearl barley; and both the London and Edinburgh make common barley an ingredient in the pectoral decoction.

216. HUMULUS Lupulus. THE HOP.—The flowers and seed-vessels are used in gout and rheumatism, under the form of infusion in boiling-water. The powder formed into an ointment with lard, is said to ease the pain of open cancer. A pillow stuffed with hops is an old and successful mode of procuring sleep in the watchfulness of delirious fever.

217. HYOSCYAMUS niger. HENBANE. Leaves and Seeds. L. E.—Henbane is a strong narcotic poison, and many instances of its deleterious effects are recorded by different authors; from which it appears, that any part of the plant, when taken in sufficient quantity, is capable of producing very dangerous and terrible symptoms. It is however much employed in the present days as an anodyne. Dr. Withering found it of great advantage in a case of difficult deglutition. Stoerck and some others recommend this extract in the dose of one grain or two; but Dr. Cullen observes, that he seldom discovered its anodyne effects till he had proceeded to doses of eight or ten grains, and sometimes to fifteen and even to twenty. The leaves of Henbane are said to have been applied externally with advantage, in the way of poultice, to resolve scirrhous tumours, and to remove some pains of the rheumatic and arthritic kind.

Similar Plants.—Verbascum Lychnites; V. nigrum.

The roots of the Henbane are to be distinguished by their very powerful and narcotic scent.

218. HYSSOPUS officinalis. HYSSOP. The Herb. L. E. D.—The leaves of Hyssop have an aromatic smell, and a warm pungent taste. Besides the general virtues of aromatics, they are particularly recommeded in humoral asthmas, coughs, and other disorders of the breast and lungs; and said to notably promote expectoration.

219. INULA Helenium. ELECAMPANE. Root. D.—Elecampane root possesses the general virtues of alexipharmics: it is principally recommended for promoting expectoration in humoural asthmas and coughs; in which intention, it used to be employed in the Edinburgh Pharmacopoeia: liberally taken, it is said to excite urine, and loosen the belly. In some parts of Germany, large quantities of this root are candied, and used as a stomachic, for strengthening the tone of the viscera in general, and for attenuating tenacious juices. Spiritous liquors extract its virtues in greater perfection than watery ones: the former scarce elevate any thing in distillation: with the latter, an essential oil arises, which concretes into white flakes; this possesses at first the flavour of the elecampane, but is very apt to lose it in keeping.

220. JUNIPERUS Sabina. SAVINE. The Tops. L. E. D.—Savine is a warm irritating aperient medicine, capable of promoting all the glandular secretions. The distilled oil is one of the most powerful emmenagogues; and is found of good service in obstructions of the uterus, or other viscra, proceeding from a laxity and weakness of the vessels, or a cold sluggish indisposition of the juices.

Similar Plants.—Juniperus oxycedrus; J. Phoenicea. These should be particularly distinguished, as Savine is attended with danger when taken immoderately.

221. JUNIPERUS communis. JUNIPER. Berries. L. E. D.—Juniper berries have a strong, not disagreeable smell; and a warm, pungent sweet taste, which, if they are long chewed, or previously well bruised, is followed by a bitterish one. The pungency seems to reside in the bark; the sweet in the juice; the aromatic flavour in oily vesicles, spread through the substance of the pulp, and distinguishable even by the eye; and the bitter in the seeds: the fresh berries yield, on expression, a rich, sweet, honey-like, aromatic juice; if previously pounded so as to break the seeds, the juice proves tart and bitter.

222. LACTUCA virosa. WILD LETTUCE. Leaves. E.—Dr. Collin at Vienna first brought the Lactuca virosa into medical repute; and its character has lately induced the College of Physicians at Edinburgh to insert it in the Catalogue of the Materia Medica. More than twenty-four cases of dropsy are said by Collin to have been successfully treated, by employing an extract prepared from the expressed juice of this plant, which is stated not only to be powerfully diuretic, but, by attenuating the viscid humours, to promote all the secretions, and to remove visceral obstructions. In the more simple cases proceeding from debility, the extract in doses of eighteen to thirty grains a-day, proved sufficient to accomplish a cure; but when the disease was inveterate, and accompanied with visceral obstructions, the quantity of extract was increased to three drams; nor did larger doses, though they excited nausea, ever produce any other bad effect; and the patients continued so strong under the use of this remedy, that it was seldom necessary to employ any tonic medicines.—Woodville's Med. Bot. p. 76.

Similar Plants.—Sonchus arvensis; Lactuca Scariola.

223. LAVANDULA Spica. LAVENDER. Flowers. L. D.—Lavender has been an officinal plant for a considerable time, though we have no certain accounts of it given by the ancients. Its medical virtue resides in the essential oil, which is supposed to be a gentle corroborant and stimulant of the aromatic kind; and is recommended in nervous debilities, and various affections proceeding from a want of energy in the animal functions.—Woodville's Med. Bot. p. 323.

224. LAURUS nobilis. BAY-TREE. Leaves and Berries. L.—In distillation with water, the leaves of bay yield a small quantity of very fragrant essential oil; with rectified spirit, they afford a moderately warm pungent extract. The berries yield a larger quantity of essential oil: they discover likewise a degree of unctuosity in the mouth; give out to the press an almost insipid fluid oil; and on being boiled in water, a thicker butyraceous one of a yellowish-green colour, impregnated with the flavour of the berry. An infusion of the leaves is sometimes drunk as tea; and the essential oil of the berries may be given from one to five or six drops on sugar, or dissolved by means of mucilages, or in spirit of wine.—Woodville's Med Bot. p. 680, 681.

225. LAURUS Sassafras. SASSAFRAS-TREE. Bark. L. E. D.—Its medical character was formerly held in great estimation; and its sensible qualities, which are stronger than any of the woods, may have probably contributed to establish the opinion so generally entertained of its utility in many inveterate diseases: for, soon after its introduction into Europe, it was sold at a very high price, and its virtues were extolled in publications professedly written on the subject. It is now, however, thought to be of very little importance, and seldom employed but in conjunction with other medicines of a more powerful nature.

Dr. Cullen found that a watery infusion of it taken warm and pretty largely, was very effectual in promoting sweat; but he adds, "to what particular purpose this sweating was applicable, I have not been able to determine." In some constitutions sassafras, by its extreme fragrance, is said to produce headache: to deprive it of this effect, the decoction ought to be employed.—Woodville's Mat. Med. p. 677.

226. LEONTODON Taraxicum. N EBION. Root. L.—The roots contain a bitter milky juice; they promise to be of use as asperient and detergent medicines; and have sometimes been directed in this intention with good success. Boerhaave esteems them capable, if duly continued, of resolving almost all kinds of coagulations, and opening very obstinate obstructions of the viscera.

227. LINUM usitatissimum. FLAX. The Seeds. L. E.— Linseed yields to the press a considerable quantity of oil; and boiled in water, a strong mucilage: these are occasionally made use of for the same purposes as other substances of that class; and sometimes the seeds themselves in emollient and maturating cataplasms. They have also been employed in Asia, and, in times of scarcity, in Europe, as food: but are not agreeable, or in general wholesome.

228. LINUM catharticum. PURGING-FLAX. The Herb. L. D.-This is a very small plant, not above four or five inches high, found wild upon chalky hills, and in dry pasture-grounds. Its virtue is expressed in its title: an infusion in water or whey of a handful of the fresh leaves, or a dram of them in substance when dried, is said to purge without inconvenience.

229. LOBELIA siphylitica. BLUE CARDINAL FLOWER. The Root. E.—Every part of the plant abounds with a milky

juice, and has a rank smell. The root, which is the part directed for medicinal use, in taste resembles tobacco, and is apt to excite vomiting. It derived its name, Siphylitica, from its efficacy in the cure of Siphylis, as experienced by the North American Indians, who considered it a specific to that disease.

A decoction was made of a handful of the roots in three measures of water. Of this, half a measure is taken in the morning fasting, and repeated in the evening; and the dose is gradually increased till its purgative effects become too violent, when the decoction is to be intermitted for a day or two, and then renewed till a perfect cure is effected. But it does not appear that the antisiphylitic powers of Lobelia have been confirmed by any instances of European practice.— Woodville's Med. Bot. p. 251.

230. LYTHRUM Salicaria. WILLOW HERB. The Herb. D.—This is used internally in dropsies, obstinate gleets, and leucorrhoea.

Similar Plants.—Epilobium palustre; Epilob. angustifolium; Epilob. hirsutum.

231. MALVA sylvestris. COMMON MALLOW. Herb. L. E.—The leaves are ranked the first of the four emollient herbs: they were formerly of some esteem, in food, for loosening the belly; at present, decoctions of them are sometimes employed in dysenteries, heat and sharpness of urine, and in general for obtunding acrimonious humours: their principal use is in emollient glysters, cataplasms, and fomentations.

232. MARRUBIUM vulgare. HORFHOUND. Herb. E. D.—It is greatly extolled for its efficacy in removing obstructions of the lungs and other viscera. It has chiefly

been employed in humoural asthmas. Mention is made of its successful use in scirrhous affections of the liver, jaundice, cachexies, and menstrual suppressions.—Woodville's Med. Bot. p. 333.

Similar Plants.—Ballota nigra; B. alba.

233. MELISSA officinalis. BALM. Herb. L. E.—This herb, in its recent state, has a weak roughish aromatic taste, and a pleasant smell, somewhat of the lemon kind. On distilling the fresh herb with water, it impregnates the first runnings pretty strongly with its grateful flavour. Prepared as tea, however, it makes a grateful diluent drink in fevers; and in this way it is commonly used, either by itself, or acidulated with the juice of lemons.—Woodville's Med. Bot. p. 335, 336.

234. MENTHA viridis. SPEAR-MINT. Leaves. L. D.—The virtues of Mint are those of a warm stomachic and carminative: in loss of appetite, nauseae, continual retchings to vomit, and (as Boerhaave expresses it) almost paralytic weakness of the stomach, there are few simples perhaps of equal efficacy. In colicky pains, the gripes to which children are subject, lienteries, and other kinds of immoderate fluxes, this plant frequently does good service. It likewise proves beneficial in sundry hysteric cases, and affords an useful cordial in languors and other weaknesses consequent upon delivery. The best preparations for these purposes are, a strong infusion made from the dry leaves in water (which is much superior to one from the green herb) or rather a tincture or extract prepared with rectified spirit.

The essential oil, a simple and spirituous water, and a conserve, are kept in the shops: the Edinburgh College directs an infusion of the leaves in the distilled water. This herb is an ingredient also in the three alexitereal waters;

and its essential oil in the stomach plaster and stomach pills.—Lewis's Mat. Med.

235. MENTHA Piperita. PEPPER-MINT. Herb. L. E. D.—The leaves have a more penetrating smell than any of the other mints, and a much warmer, pungent, glowing taste like pepper, sinking as it were into the tongue. The principal use of this herb is in flatulent colics, languors, and other like disorders; it seems to act as soon as taken, and extends its effects through the whole system, instantly communicating a glowing warmth. Water extracts the whole of the pungency of this herb by infusion, and elevates it in distillation. Its officinal preparations are an essential oil, and a simple and spirituous water.

236. MENTHA Pulegium. PENNYROYAL. Herb. L. E. D.—Pennyroyal is a warm pungent herb of the aromatic kind, similar to mint, but more acrid and less agreeable. It has long been held in great esteem, and not undeservedly, as an aperient and deobstruent, particularly in hysteric complaints, and suppressions of the uterine purgations. For these purposes, the distilled water is generally made use of, or, what is of equal efficacy, an infusion of the leaves. It is observable, that both water and rectified spirit extract the virtues of this herb by infusion, and likewise elevate greatest part of them in distillation.—Lewis's Mat. Med.

237. MENYANTHES trifoliata. BUCK-BEAN. Leaves. L. E. D.—This is an efficacious aperient and deobstruent; it promotes the fluid secretions, and, if liberally taken, gently loosens the belly. It has of late gained great reputation in scorbutic and scrophulous disorders; and its good effects in these cases have been warranted by experience: inveterate cutaneous diseases have been removed by an infusion of the leaves, drunk to the quantity of a pint a-day, at proper intervals, and continued some weeks. Boerhaave relates,

that he was relieved of the gout by drinking the juice mixed with whey.

238. MOMORDICA Elaterium. SPIRTING CUCUMBER. Fruit L. E. D.—Elaterium is a strong cathartic, and very often operates also upwards. Two or three grains are accounted in most cases a sufficient dose. Simon Paulli relates some instances of the good effects of this purgative in dropsies: but cautions practitioners not to have recourse to it till after milder medicines have proved ineffectual; to which caution we heartily subscribe. Medicines indeed in general, which act with violence in a small dose, require the utmost skill to manage them with any tolerable degree of safety: to which may be added, that the various manners of making these kinds of preparations, as practised by different hands, must needs vary their power.

239. MORUS nigra. MULBERRY. Fruit. L.—It has the common qualities of the other sweet fruits, abating heat, quenching thirst, and promoting the grosser secretions; an agreeable syrup made from the juice is kept in the shops. The bark of the roots has been in considerable esteem as a vermifuge; its taste is bitter, and somewhat astringent.— Lewis's Mat. Med.

240. NICOTIANA Tabacum. TOBACCO. Leaves. L. E. D.—Tobacco is sometimes used externally in unguents for destroying cutaneous insects, cleansing old ulcers, &c. Beaten into a mash with vinegar or brandy, it has sometimes proved serviceable for removing hard tumours of the hypochondres.

241. ORIGANUM Majorana. SWEET MARJORAM. Herb. L. E.-It is a moderately warm aromatic, yielding its virtues both to aqueous and spirituous liquors by infusion, and to water in distillation. It is principally celebrated in

disorders of old people. An essential oil of the herb is kept in the shops. The powder of the leaves proves an agreeable errhine.

242. ORIGANUM vulgare. POT MARJORAM. Herb. L. D.—It has an agreeable aromatic smell approaching to that of marjoram, and a pungent taste much resembling thyme, to which it is likewise thought to be more nearly allied in its medicinal qualities than to any of the other verticillatae, and therefore deemed to be emmenagogue, tonic, stomachic, &c.

The dried leaves used instead of tea are said to be extremely grateful. They are also employed in medicated baths and fomentations.—Woodville's Med. Bot. p. 345.

243. OXALIS Acetosella. WOOD SORREL. Herb. L.—In taste and medical qualities it is similar to the common sorrel, but considerably more grateful, and hence is preferred by the London College. Boiled with milk, it forms an agreeable whey; and beaten with sugar, a very elegant conserve.—Lewis's Mat. Med.

244. PAPAVER Rhoeas. RED POPPY. Petals. L. E. D.—The flowers of this plant yield upon expression a deep red juice, and impart the same colour by infusion to aqueous liquors. A syrup of them is kept in the shops: this is valued chiefly for its colour; though some expect from it a lightly anodyne virtue.

245. PAPAVER somniferum. OPIUM POPPY. Gum. L. E. D.-Poppy heads, boiled in water, impart to the menstruum their narcotic juice, together with the other juices which they have in common with vegetable matters in general. The liquor strongly pressed out, suffered to settle, clarified with whites of eggs, and evaporated to a due consistence,

yields about one-fifth or one-sixth the weight of the heads, of extract. This possesses the virtues of opium; but requires to be given in double its dose to answer the same intention, which it is said to perform without occasioning nausea and giddiness, the usual consequences of the other.

The general effects of this medicine are, to relax the solids, ease pain, procure sleep, promote perspiration, but restrain all other evacuations. When its operation is over, the pain, and other symptoms which it had for a time abated, return; and generally with greater violence than before, unless the cause has been removed by the diaphoresis or relaxation which it occasioned.

The operation of opium is generally attended with a slow, but strong and full pulse, a dryness of the mouth, a redness and light itching of the skin: and followed by a degree of nausea, a difficulty of respiration, lowness of the spirits, and a weak languid pulse.

With regard to the dose of opium, one grain is generally sufficient, and often too large a one; maniacal persons, and those who have been long accustomed to take it, require three or more grains to have the due effect. Among the eastern nations, who are habituated to opium, a dram is but a moderate dose: Garcias relates, that he knew one who every day took ten drams. Those who have been long accustomed to its use, upon leaving it off, are seized with great lowness, languor, and anxiety; which are relieved by having again recourse to opium, and, in some measure, by wine or spirituous liquors.

Similar Plants.—Papaver hybridum; P. Argemone.

246. PASTINACA Opoponax. OPOPONAX, or CANDY CARROT. Gum Opoponax. L.— The juice is brought from

Turkey and the East Indies, sometimes in round drops or tears, but more commonly in irregular lumps, of a reddish-yellow colour on the outside, with specks of white, inwardly of a paler colour, and frequently variegated with large white pieces.

Boerhaave frequently employed it, along with ammoniacum and galbanum, in hypochondriacal disorders, obstructions of the abdominal viscera from a sluggishness of mucous humours, and a want of due elasticity of the solids.

247. PIMPINELLA Anisum. ANISEED. The Seeds. L. E. D.-These seeds are in the number of the four greater hot seeds: their principal use is in cold flatulent disorders, where tenacious phlegm abounds, and in the gripes to which young children are subject. Frederick Hoffman strongly recommends them in weakness of the stomach, diarrhoeas, and for strengthening the tone of the viscera in general; and thinks they well deserve the appellation given them by Helmont, intestinorum solamen.

248. PINUS sylvestris. SCOTCH FIR. Tar, yellow Resin, and Turpentine. L. D.—Tar, which is well known from its oeconomical uses, is properly an empyreumatic oil of turpentine, and has been much used as a medicine, both internally and externally. Tar-water, or water impregnated with the more soluble parts of tar, was some time ago a very popular remedy in various obstinate disorders, both acute and chronic, especially in small-pox, scurvy, ulcers, fistulas, rheumatisms, &c.

Turpentine is an extract also from the same tree, which is used for various purposes of medicine and the arts.

249. PINUS Abies. SPRUCE-FIR. Burgundy Pitch. L. E. D.—This is entirely confined to external use, and was formerly an ingredient in several ointments and plasters. In inveterate coughs, affections of the lungs, and other internal complaints, plasters of this resin, by acting as a tropical stimulus, are frequently found of considerable service.—Woodville's Med. Bot.

250. POLYGONUM Bistorta. BISTORT. The Roots. L. E. D.—All the parts of bistort have a rough austere taste, particularly the root, which is one of the strongest of the vegetable stringents. It is employed in all kinds of immoderate haemorrhages and other fluxes, both internally and externally, where astringency is the only intention. It is certainly a very powerful styptic, and is to be looked on simply as such; the sudorific, antipestilential, and other like virtues attributed to it, it has no other claim to, than in consequence of this property, and of the antiseptic power which it has in common with other vegetable styptics. The largest dose of the root in powder is one dram.

251. PRUNUS domestica. FRENCH PRUNES. The Fruit. L. E. D.—The medical effects of the damson and common prunes are, to abate heat, and gently loosen the belly: which they perform by lubricating the passage, and softening the excrement. They are of considerable service in costiveness accompanied with heat or irritation, which the more stimulating cathartics would tend to aggravate: where prunes are not of themselves sufficient, their effects may be promoted by joining with them a little rhubarb or the like; to which may be added some carminative ingredient, to prevent their occasioning flatulencies. Prunelloes have scarce any laxative quality: these are mild grateful refrigerants, and, by being occasionally kept in the mouth, usefully allay the thirst of hydropic persons.

252. PUNICA Granatum. POMEGRANATE. Rind of the Fuit. L. E. D.—This fruit has the general qualities of the other sweet summer fruits, allaying heat, quenching thirst, and gently loosening the belly. The rind is a strong astringent, and as such is occasionally made use of.

253. PYRUS Cydonia. QUINCE. The Kernels. L.—The seeds abound with a mucilaginous substance, of no particular taste, which they readily impart to watery liquors: an ounce will render three pints of water thick and ropy like the white of an egg. A syrup and jelly of the fruit, and mucilage of the seeds, used to be kept in the shops.

254. QUEROUS pedunculata. OAK. Bark. L. E. D.—This bark is a strong astringent; and hence stands recommended in haemorrhagies, alvine fluxes, and other preternatural or immoderate secretions.

255. RHAMNUS catharticus. BUCKTHORN. Berries. L. E.—Buckthorn-berries have a faint disagreeable smell, and a nauseous bitter taste. They have long been in considerable esteem as cathartics; and celebrated in dropsies, rheumatisms, and even in the gout; though in these cases they have no advantage above other purgatives, and are more offensive, and operate more churlishly, than many which the shops are furnished with: they generally occasion gripes, sickness, dry the mouth and throat, and leave a thirst of long duration. The dose is about twenty of the fresh berries in substance, and twice or thrice this number in decoction, an ounce of the expressed juice, or a dram of the dried berries.

256. RHEUM palmatum. TURKEY RHUBARB. Roots. L. E. D.—Rhubarb is a mild cathartic, which operates without violence or irritation, and may be given with safety even to pregnant women and to children. In some people, however,

it always occasions severe griping. Besides its purgative quality, it is celebrated for an astringent one, by which it strengthens the tone of the stomach and intestines, and proves useful in diarrhoea and disorders proceeding from a laxity of the fibres. Rhubarb in substance operates more powerfully as a cathartic than any of the preparations of it. Watery tinctures purge more than the spirituous ones; whilst the latter contain in greater perfection the aromatic, astringent, and corroborating virtues of the rhubarb. The dose, when intended as a purgative, is from a scruple to a dram or more.

The Turkey rhubarb is, among us, universally preferred to the East India sort.

The plant is common in our gardens, but their medicinal powers are much weaker than in those from abroad.

RHODODENDRON Chrysanthemum. YELLOW-FLOWERED RHODODENDRON. See No. 290.

257. RHUS Toxicodendron. POISON-OAK. Leaves. L. E.—Of considerable use in paralytic affections, and is much used in the present day.

It is, however, often substituted by the Rhus radicans, which has not the medical properties that this plant has; and it is to be regretted that the leaves of both species are so much alike, that, when gathered, they are not to be distinguished.

258. RICINUS communis. PALMA CHRISTI. Seeds and Oil. L. E. D.—The oil, commonly called nut or castor oil, is got by expression, retains somewhat of the mawkishness and acrimony of the nut; but is, in general, a safe and mild laxative in cases where we wish to avoid irritation, as in

those of colic, calculus, gonorrhoea, &c. and some likewise use it as a purgative in worm-cases. Half an ounce or an ounce commonly answers with an adult, and a dram or two with an infant. The castor oil which is imported is not so good as the expressed oil from the nut made in this country. The disagreeable taste is from the coats of the seeds; the best kind is pressed out after the seeds are decorticated.

259. ROSA centifolia. DAMASK ROSE. Petals. L. E. D.— In distillation with water, it yields a small portion of a butyraceous oil, whose flavour exactly resembles that of the roses. This oil, and the distilled water, are very useful and agreeable cordials. Hoffmann strongly recommends them as of singular efficacy for raising the strength, cheering and recruiting the spirits, and allaying pain; which they perform without raising any heat in the constitution, rather abating it when inordinate. Although the damask rose is recommended by Dr. Woodville, yet, having grown this article for sale, I find that the preference is always given to the Provence rose by those who distil them.

260. ROSA gallica. RED OFFICINAL ROSE. Petals. L. E. D.-This has very little of the fragrance of the foregoing sort; it is a mild and grateful astringent, especially before the flower has opened: this is considerably improved by hasty exsiccation, but both the astringency and colour are impaired by slow drying. In the shops are prepared a conserve and a tincture.

261. ROSA canina. DOG-ROSE. The Pulp of the Fruit. L. E.-The fruit, called heps or hips, has a sourish taste, and obtains a place in the London Pharmacopoeia in the form of a conserve: for this purpose, the seeds and chaffy fibres are to be carefully removed; for, if these prickly fibres are not entirely scraped off from the internal surface of the hips,

the conserve is liable to produce considerable irritation on the primae viae.

262. ROSMARINUS officinalis. ROSEMARY. Tops. L. E. D.—Rosemary has a fragrant smell and a warm pungent bitterish taste, approaching to those of lavender: the leaves and tender tops are strongest; next to these the cup of the flower; the flowers themselves are considerably the weakest, but most pleasant. Aqueous liquors extract great share of the virtues of rosemary leaves by infusion, and elevate them in distillation: along with the water arises a considerable quantity of essential oil, of an agreeable strong penetrating smell. Pure spirit extracts in great perfection the whole aromatic flavour of the rosemary, and elevates very little of it in distillation: hence the resinous mass left upon abstracting the spirit, proves an elegant aromatic, very rich in the peculiar qualities of the plant. The flowers of rosemary give over great part of their flavour in distillation with pure spirit; by watery liquors, their fragrance is much injured; by beating, destroyed.

263. RUBIA tinctorum. MADDER. Roots. L. E. D.—It has little or no smell; a sweetish taste, mixed with a little bitterness. The virtues attributed to it are those of a detergent and aperient; whence it has been usually ranked among the opening roots, and recommended in obstructions of the viscera, particularly of the kidneys, in coagulations of the blood from falls or bruises, in the jaundice, and beginning dropsies.

It is observable, that this root, taken internally, tinges the urine of a deep red colour; and in the Philosophical Transactions we have an account of its producing a like effect upon the bones of animals which had it mixed with their food: all the bones, particularly the more solid ones, were changed, both externally and internally, to a deep red, but neither the fleshy nor cartilaginous parts suffered any alteration: some of these bones macerated in water for many weeks together, and afterwards steeped and boiled in spirit of wine, lost none of their colour, nor communicated any tinge to the liquors.

264. RUMEX Acetosa. SORREL. Leaves. L.—These have an agreeable acid taste. They have the same medicinal qualities as the Oxalis Acetosella, and are employed for the same purposes.

Sorrel taken in considerable quantities, or used prepared for food, will be found of great advantage when a refrigerant and antiscorbutic regimen is required.—Woodville's Med. Bot.

265. RUTA graveolens. RUE. Leaves. L. E. D.—These are powerfully stimulating, attenuating, and detergent: and hence, in cold phlegmatic habits, they quicken the circulation, dissolve tenacious juices, open obstructions of the excretory glands, and promote the fluid secretions. The writers on the Materia Medica in general have entertained a very high opinion of the virtues of this pant. Boerhaave is full of its praises; particularly of the essential oil, and the distilled water cohobated or redistilled several times from fresh parcels of the herb: after somewhat extravagantly commending other waters prepared in this manner, he adds, with regard to that of rue, that the greatest commendations he can bestow upon it fall short of its merit: "What medicine (says he) can be more efficacious for promoting perspiration, in cases of epilepsies, and for expelling poison?" Whatever service rue may be of generally, it undoubtedly has its use in the two last cases: the cohobated water, however, is not the most efficacious preparation.

266. SALIX fragilis. CRACK WILLOW. Bark. L. D.-The bark of the branches of this tree manifests a considerable degree of bitterness to the taste, and is also astringent; hence it has been thought a good substitute for the Peruvian bark, and, upon trial, was found to stop the paroxysms of intermittents: it is likewise recommended in other cases requiring tonic or astringent remedies. Not only the bark of this species of Salix, but that of several others, possess similar qualities, particularly of the Salix alba pentandria, and capraea, all of which are recommended in foreign Pharmacopoeias. But, in our opinion, the bark of the Salix triandria is more effectual than that of any other of this genus; at least, its sensible qualities give it a decided preference.—Woodville's Med Bot.

267. SALVIA officinalis. GREEN AND RED SAGE. Herb. E. D.—Its effects are, to moderately warm and strengthen the vessels; and hence, in cold phlegmatic habits, it excites appetite, and proves serviceable in debilities of the nervous system.

The red sage, mixed with honey and vinegar, is used for a gargle in sore throats. Aqueous infusions of the leaves, with the addition of a little lemon juice, prove an useful diluting drink in febrile disorders, of an elegant colour, and sufficiently acceptable to the palate.

268. SAMBUCUS nigra. COMMON ELDER. Flowers and Berries. L. E. D.—The parts of the Sambucus which are proposed for medicinal use in the Pharmacopoeias, are the inner bark, the flowers, and the berries. The flowers have an agreeable flavour, which they give over in distillation with water, and impart by infusion, both to water and rectified spirit: on distilling a large quantitiy of them with water, a small portion of a butyraceous essential oil separates. Infusions made from the fresh flowers are gently

laxative and aperient; when dry, they are said to promote chiefly the cuticular excretion, and to be particularly serviceable in erysipetalous and eruptive disorders.—Woodville's Med. Bot. 598.

269. SCILLA maritima. SQUILL. Root. L. E. D.—This root is to the taste very nauseous, intensely bitter and acrimonious; much handled, it exulcerates the skin. With regard to its medical virtues, it powerfully stimulates the solids, and attenuates viscid juices; and by these qualities promotes expectoration, urine, and perspiration: if the dose is considerable, it proves emetic, and sometimes purgative. The principal use of this medicine is where the primae viae abound with mucous matter, and the lungs are oppressed by tenacious phlegm.

270. SCROPHULARIA nodosa. KNOTTY FIGWORT. Herb. D.—The roots are of a white colour, full of little knobs or protuberances on the surface: this appearance gained it formerly some repute against scrophulous disorders and the piles; and from hence it received its name: but modern practitioners expect no such virtues from it. It has a faint unpleasant smell, and a somewhat bitter disagreeable taste.

271. SINAPIS nigra. BLACK MUSTARD. Seeds. L. E. D.—By writers on the Materia Medica, mustard is considered to promote appetite, assist digestion, attenuate viscid juices, and, by stimulating the fibres, to prove a general remedy in paralytic and rheumatic affections. Joined to its stimulant qualities, it frequently, if taken in considerable quantity, opens the body, and increases the urinary discharge; and hence has been found useful in dropsical complaints.—Woodville's Med. Bot. p. 404.

272. SINAPIS alba. WHITE MUSTARD. Seeds. L. E. D.—These have been recommended to be taken whole in cases of rheumatism and have been known to produce considerable relief.

273. SISYMBRIUM Nasturtium. WATER-CRESSES. Herb. E.-Hoffman recommends this as of singular efficacy for accelerating the circulation, strengthening the viscera, opening obstructions of the glands, promoting the fluid secretions, and purifying the blood and humours: for these purposes, the expressed juice, which contains the peculiar taste and pungency of the herb, may be taken in doses of an ounce or two, and continued for a considerable time.

274. SIUM nodiflorum. CREEPING WATER-PARSNEP. The Root. D.-This plant has not been admitted into the Materia Medica of any of the Pharmacopoeias which we have seen, except that of the London College, into which it was received in the character of an antiscorbutic, or rather as the corrector of acrid humours, especially when manifested by cutaneous eruptions and tumours in the lymphatic system, for which we have the testimony of Beirie and Ray; but the best proofs of its efficacy are the following given by Dr. Withering: "A young lady, six years old, was cured of an obstinate disease by taking three large spoonfuls of the juice twice-a-day; and I have repeatedly given to adults three or four ounces every morning in similar complaints with the greatest advantage. It is not nauseous; and children take it readily if mixed with milk. In the dose I have given, it neither affects the head, the stomach, nor the bowels." Woodville's Med. Bot. 146.

275. SMILAX Sarsaparilla. SARSAPARILLA. Root. L. E. D.—This root was first brought into Europe by the Spaniards, about the year 1565, with the character of a specific for the cure of the lues venerea, which made its

appearance a little before that time, and likewise of several obstinate chronic disorders. Whatever good effects it might have produced in the warmer climates, it proved unsuccessful in this. It appears, however, from experience, that though greatly unequal to the character which it bore at first, it is in some cases of considerable use as a sudorific, where more acrid medicines are improper.

276. SOLANUM Dulcamara. BITTERSWEET. Stalk. L. D.—The taste of the twigs and roots, as the name of the plant expresses, is both bitter and sweet; the bitterness being first perceived, and the sweet afterwards. They are commended for resolving coagulated blood, and as a cathartic, diuretic, and deobstruent.

277. SOLIDAGO Virga aurea. GOLDEN ROD. Flowers and Leaves. D.—The leaves have a moderately astringent bitter taste, and hence prove serviceable in debility and laxity of the viscera, and disorders proceeding from that cause.

278. SPARTIUM scoparium. BROOM. Tops and Seeds. L. D.-These have a nauseous bitter taste: decoctions of them loosen the belly, promote urine, and stand recommended in hydropic cases. The flowers are said to prove cathartic in decoction, and emetic in substance, though in some places, as Lobel informs us, they are commonly used, and in large quantity, in salads, without producing any effect of this kind. The qualities of the seeds are little better determined: some report that they purge almost as strongly as hellebore, in the dose of a dram and a half; whilst the author above mentioned relates, that he has given a decoction of two ounces of them as a gentle emetic.

279. SPIGELLA marylandica. WORM GRASS. Root. L. E. D.-About forty years ago, the anthelmintic virtues of the

root of this plant were discovered by the Indians; since which time it has been much used here. I have given it in hundreds of cases, and have been very attentive to its effects. I never found it do much service, except when it proved gently purgative. Its purgative quality naturally led me to give it in febrile diseases which seem to arise from viscidity in the primae viae; and in these cases it succeeded to admiration, even when the sick did not void worms.

To a child of two years of age who had been taking ten grains of the root twice a-day without having any other effect than making her dull and giddy, I prescribed twenty-two grains morning and evening, which purged her briskly, and brought away five large worms. [Communications from Dr. Gardner.]-Woodville's Med. Bot.

280. TANACETUM vulgare. TANSY. Herb. E. D.—Considered as a medicine, it is a moderately warm bitter, accompanied with a strong, not very disagreeable flavour. Some have had a great opinion of it in hysteric disorders, particularly those proceeding from a deficiency or suppression of the usual course of nature.

281. TEUCRIUM Marum. CAT THYME. Herb. D.—The leaves have an aromatic bitterish taste; and, when rubbed betwixt the fingers, a quick pungent smell, which soon affects the head, and occasions sneezing: distilled with water, they yield a very acrid, penetrating essential oil, resembling one obtained by the same means from scurvy-grass. These qualities sufficiently point out the uses to which this plant might be applied; at present, it is little otherwise employed than in cephalic snuffs.

282. TEUCRIUM Chamaedrys. GERMANDER. Herb. D.—The leaves, tops, and seeds, have a bitter taste, with some degree of astringency and aromatic flavour. They

were recommended as sudorific, diuretic, and emmenagogue, and for strengthening the stomach and viscera in general. With some they have been in great esteem in intermittent fevers; as also in scrophulous and other chronic disorders.

283. TORMENTILLA erecta. TORMENTIL, or UPRIGHT SEPTFOIL. Root. L. E. D. —The root is the only part of this plant which is used medicinally; it has a strong styptic taste, but imparts no peculiar sapid flavour. This has been long held in great estimation as an astringent. Dr. Cullen has used it with gentian with great effect in intermittent fevers. Lewis recommends an ounce and a half of the powdered root to be boiled in three pints of water to a quart, adding towards the end of the boiling a dram of cinnamon. Of the strained liquor, sweetened with an ounce of any agreeable syrup, two ounces or more may be taken four or five times a-day.

284. TUSSILAGO Farfara. COLTSFOOT. Herb. L. E. D.—Tussilago stands recommended in coughs and other disorders of the breast and lungs: the flowers were an ingredient in the pectoral decoction of the Edinburgh Pharmacopoeia.

285. VALERIANA officinalis. VALERIAN. Root. L. E. D.—Valerian is a medicine of great use in nervous disorders, and is particularly serviceable in epilepsies proceeding from a debility in the nervous system. It was first brought into esteem in these cases by Fabius Columna, who by taking the powdered root, in the dose of half a spoonful, was cured of an inveterate epilepsy after many other medicines had been tried in vain. Repeated experience has since confirmed its efficacy in this disorder; and the present practice lays considerable stress upon it.

286. VERATRUM album. WHITE HELLEBORE. Root. L. E. D.-The root has a nauseous, bitterish, acrid taste, burning the mouth and fauces: wounded when fresh, it emits an extremely acrimonious juice, which mixed with the blood, by a wound, is said to prove very dangerous: the powder of the dry root, applied to an issue, occasions violent purging: snuffed up the nose, it proves a strong, and not always a safe, sternutatory. This root, taken internally, acts with extreme violence as an emetic, and has been observed, even in a small dose, to occasion convulsions and other terrible disorders. The ancients sometimes employed it in very obstinate cases, and always made this their last resource.

Similar Plant.—Gentiana lutea, which see.

287. VERONICA Beccabunga. BROOKLIME. Herb. L. D.—This plant was formerly considered of great use in several diseases, and was applied externally to wounds and ulcers; but if it have any peculiar efficacy, it is to be derived from its antiscorbutic virtue.

As a mild refrigerant juice, it is preferred where an acrimonious state of the fluids prevails, indicated by prurient eruptions upon the skin, or in what has been called the hot scurvy.—Woodville's Med. Bot. 364.

288. VITIS vinifera. GRAPE VINE. Raisins and different Wines. L. E.— These are to cheer the spirits, warm the habit, promote perspiration, render the vessels full and turgid, raise the pulse, and quicken the circulation. The effects of the full-bodied wines are much more durable than those of the thinner; all sweet wines, as Canary, abound with a glutinous nutritious substance; whilst the others are not nutrimental, or only accidentally so by strengthening the organs employed in digestion: sweet wines in general

do not pass off freely by urine, and heat the constitution more than an equal quantity of any other, though containing full as much spirit: red port, and most of the red wines, have an astringent quality, by which they strengthen the tone of the stomach and intestines, and thus prove serviceable for restraining immoderate secretions: those which are of an acid nature, as Renish, pass freely by the kidneys, and gently loosen the belly: it is supposed that these last exasperate, or occasion gout and calculous disorders, and that new wines of every kind have this effect.

The ripe fruit of grapes, of which there are several kinds, properly cured and dried, are the raisins and currants of the shops: the juice of these also, by fermentation, affords wine as well as vinegar and tartar.

The medical use of raisins is, their imparting a very pleasant flavour both to aqueous and spiritous menstrua. The seeds or stones are supposed to give a disagreeable relish, and hence are generally directed to be taken out: nevertheless I have not found that they have any disagreeable taste.—Lewis's Mat. Med.

289. ULMUS campestris. ELM. Bark. L. E. D.—The leaves have a bitterish astringent taste, and are recommended in powder, to the extent of at least two drams a-day, in ulcerations of the urinary passages and catarrhus vesicae. The powder has been used with opium, the latter being gradually increased to a considerable quantity, in diabetes, and it is said with advantage. Some use it for alleviating the dyspeptic symptoms in nephritic calculous ailments.—Lewis's Mat. Med.

290. RHODODENDRON Chrysanthemum. YELLOW-FLOWERED RHO-DODENDRON. E. The Leaves.—This

species of Rhododendron has lately been introduced into Britain: it is a native of Siberia, affecting mountainous situations, and flowering in June and July.

Little attention was paid to this remedy till the year 1779, when it was strongly recommended by Koelpin as an efficacious medicine, not only in rheumatism and gout, but even in venereal cases; and it is now very generally employed in chronic rheumatisms in various parts of Europe. The leaves, which are the part directed for medicinal use, have a bitterish subastringent taste, and, as well as the bark and young branches, manifest a degree of acrimony. Taken in large doses they prove a narcotic poison, producing those symptoms which we have described as occasioned by many of the order Solanaceae.

Dr. Home, who tried it unsuccessfully in some cases of acute rheumatism, says, it appears to be one of the most powerful sedatives which we have, as in most of the trials it made the pulse remarkably slow, and, in one patient, reduced it 38 beats. And in other cases in which the Rhododendron has been used at Edinburgh, it has been productive of good effects; and, accordingly, it is now introduced into the Edinburgh Pharmacopoeia.

The manner of using this plant by the Siberians was, by putting two drams of the dried leaves in an earthen-pot with about ten ounces of boiling-water, keeping it near a boiling heat for a night, and this they took in the morning; and by repeating it three or four times it generally affected a cure. It is said to occasion heat, thirst, a degree of delirium, and a peculiar sensation of the parts affected.— Woodville's Med. Bot. p. 239.

SECT. VIII.—MEDICINAL PLANTS not contained in either of the BRITISH DISPENSATORIES.

For the use of the Medical Student I selected in the foregoing section such plants as are contained in the Pharmacopoeias of the present day: but there are many mentioned in Woodville's Medical Botany, Lewis's Dispensatory, &c. which, although discarded from the College list, are nevertheless still used by medical practitioners and others.

It would be difficult to give a full history of all the plants that have from time to time been recommended for medical uses. The old writers, as Gerard, Parkinson, Lyte, &c. attributed medical virtues to all the plants which came under their notice; and, on the other hand, as we observed above, the vegetable department of the Pharmacopoeias has from time to time been reduced so much, that, if we had confined ourselves to that alone, we fear our little treatise on this head would, by many persons, be thought defective. The following list is therefore given, as containing what are used, though probably not so much by practitioners in medicine, as by our good housewives in the country, who, without disparagement to medical science, often relieve the distresses of their families and neighbours by the judicious application of drugs of this nature, and many of which are also sold for the same purposes in the London herb-shops.

291. ACANTHUS mollis. SMOOTH BEARS-BREECH. The Leaves.—Are of a soft sweetish taste, and abound with a mucilaginous juice: its virtues do not seem to differ from those of Althea and other mucilaginous plants.

292. ACHILLA Ptarmica. SNEEZEWORT. The Root.—The roots have and acrid smell, and a hot biting taste: chewed, they occasion a plentiful discharge of saliva; and when powdered and snuffed up the nose, provoke sneezing. These are sold at the herb-shops as a substitute for pellitory of Spain.

293. ACHILLEA Ageratum. MAUDLIN. The Leaves and Flowers.—This has a light agreeable smell; and a roughish, somewhat warm and bitterish taste. These qualities point out its use as a mild corroborant; but it has long been a stranger in practice, and is now omitted both by the London and Edinburgh Colleges. It is however in use by the common people.

294. ACHILLEA Millefolium. YARROW. The Leaves.—The leaves have a rough bitterish taste, and a faint aromatic smell. Their virtues are those of a very mild astringent, and as such they stand recommended in haemorrhages both internal and external, diarrhoeas, debility and laxity of the fibres; and likewise in spasmodic hysterical affections.

295. AJUGA reptans. BUGLE. The Leaves.—These have at first a sweetish taste, which gradually becomes bitterish and roughish. They are recommended as vulnerary medicines, and in all cases where mild astringents or corroborants are proper.

296. ALCHEMILLA vulgaris. LADY'S MANTLE. The Leaves.—These discover to the taste a moderate astringency, and were formerly much esteemed in some female weaknesses, and in fluxes of the belly. They are now rarely made use of; though both the fresh leaves and roots might doubtless be of service in cases where mild astringents are required.

297. AMMI majus. BISHOPS-WEED. The Seeds.—The seeds of common bishops-weed are large and pale-coloured: their smell and taste are weak, and without any thing of the origanum flavour of the true ammi, which does not grow in this country. They are ranked among the four lesser hot seeds, but are scarcely otherwise made use of than as an ingredient in the theriaca.—Lewis's Mat. Med.

298. AMYGDALUS Persica. ALMONDS. Flowers.—They have a cathartic effect, and especially to children have been successfully given in the character of a vermifuge for this purpose; an infusion of a dram of the flowers dried, or half an ounce in their recent state, is the requisite dose. The expressed oil of almonds has been for a long time, and is at present, in use for many purposes in medicine. The concentrated acid of the bitter almond is a most dangerous poison to man and all other animals.

299. ANAGALLIS arvensis. PIMPERNEL. The Leaves.—Many extraordinary virtues have been attributed to them. Geoffroy esteems them cephalic, sudorific, vulnerary, anti-maniacal, anti-epileptic, and alexiteral.

300. ANCHUSA angustifolia. BUGLOSS. The Roots, Leaves, and Flowers.— Bugloss has a slimy sweetish taste, accompanied with a kind of coolness: the roots are the most glutinous, and the flowers the least so. These qualities point out its use in hot bilious or inflammatory distempers, and a thin acrimonious state of the fluids. The flowers are one of the four called cordial flowers: the only quality they have that can entitle them to this appellation, is, that they moderately cool and soften, without offending the palate or stomach; and thus in warm climates, or in hot diseases, may in some measure refresh the patient.

301. ANEMONE Hepatica. HEPATICA. The Leaves.—It is a cooling gently restringent herb; and hence recommended in a lax state of the fibres as a corroborant.

302. ANTIRRHINIUM Elatine. FLUELLIN. The Root, Bark, and Leaves.—They were formerly accounted excellent vulneraries, and of great use for cleansing and healing old ulcers and cancerous sores: some have recommended them internally in leprous and scrophulous disorders; as also in hydropic cases.

303. ANTIRRHINIUM Linaria. TOAD FLAX. The Flowers.—An infusion of them is said to be very efficacious in cutaneous disorders; and Hammerin gives an instance in which these flowers, with those of verbascum, used as tea, cured an exanthematous disorder, which had resisted various other remedies tried during the course of three years.—Woodville's Med. Bot. p. 372.

304. AQUILEGIA vulgaris. COLUMBINE. The Leaves, Flowers, and Seeds.—It has been looked upon as aperient; and was formerly in great esteem among the common people for throwing out the small-pox and measles. A distilled water, medicated vinegar, and conserve, were prepared from the flowers; but they have long given place to medicines of greater efficacy.

305. ARISTOLOCHIA longa. LONG BIRTHWORT. The Roots.—This is a tuberous root, sometimes about the size of the finger, sometimes as thick as a man's arm: great virtues used to be ascribed to this plant as a specific in most uterine obstructions and gout: the outside is of a brownish colour; the inside yellowish.

306. ARTEMISIA vulgaris. MUGWORT. The leaves.— These have a light aromatic smell, and an herbaceous

bitterish taste. They are principally celebrated as uterine and anti-hysteric: an infusion of them is sometimes drunk, either alone or in conjunction with other substances, in suppressions of immoderate fluxes. This medicine is certainly a very mild one, and considerably less hot than most others to which these virtues are attributed.

307. ASCLEPIAS Vincetoxium. SWALLOW WORT. The Root.—This root is esteemed sudorific, diuretic, and emmenagogue, and frequently employed by the French and German physicians as an alexipharmic, sometimes as a succedaneum to contrayerva; whence it has received the name of Contrayerva Germanorum. Among us it is rarely made use of.

308. ASPERULA odorata. SWEET WOODROOF. The Flowers.—It has an exceedingly pleasant smell, which is improved by moderate exsiccation; the taste is sub-saline, and somewhat austere. It imparts its flavour to vinous liquors. Asperula is supposed to attenuate viscid humours, and strengthen the tone of the bowels: it was recommended in obstructions of the liver and biliary ducts, and by some in epilepsies and palsies: modern practice has nevertheless rejected it.

309. ASPLENIUM Ceterach. SPLEENWORT.—It is recommended as a pectoral, and for promoting urine in nephritic cases. The virtue which it has been most celebrated for, is that which it has the least title to, i. e. diminish the spleen.

310. ASPLENIUM Scolophendrium. HARTS-TONGUE. The Leaves.—These have a roughish, somewhat mucilaginous taste. They are recommended in obstructions of the viscera, and for strengthening their tone; and have sometimes been made use of for these intentions, either

alone, or in conjunction with maiden-hair, or the other plants of similar properties.

311. ATROPA Mandragora. MANDRAKE. The Leaves.—The qualities of this plant are very doubtful: it has a strong disagreeable smell resembling that of the narcotic herbs, to which class it is usually referred. It has rarely been any otherwise made use of in medicine, than as an ingredient in one of the old officinal unguents. Both that composition and the plant itself are rejected from our Pharmacopoeias.

312. BALLOTA nigra. BASE HOREHOUND. The Leaves.—These are doubtless an useful aperient and deobstruent; promote the fluid secretions in general, and liberally taken loosen the belly. They are an ingredient only in the theriaca.

313. BELLIS perennis. DAISIES. The Leaves.—They have a subtile subacrid taste, and are recommended as vulneraries, and in asthmas and hectic fevers, and such disorders as are occasioned by drinking cold liquors when the body has been much heated.

214. BERBERIS vulgaris. BERBERRY. The Bark and Fruit.—The outward bark of the branches and the leaves have an astringent acid taste; the inner yellow bark, a bitter one: this last is said to be serviceable in the jaundice; and by some, to be an useful purgative.

The berries, which to the taste are gratefully acid, and moderately restringent, have been given with good success in bilious fluxes, and diseases proceeding from heat, acrimony, or thinness of the juices.

315. BETONICA officinalis. WOOD BETONY. The Leaves.—These and the flowers have an herbaceous,

roughish, somewhat bitterish taste, accompanied with a very weak aromatic flavour. This herb has long been a favourite among writers on the Materia Medica, who have not been wanting to attribute to it abundance of good qualities. Experience does not discover any other virtue in betony than that of a mild corroborant: as such, an infusion or light decoction of it may be drunk as tea, or a saturated tincture in rectified spirit given in suitable doses, in laxity and debility of the viscera, and disorders proceeding from thence.

316. BETULA alba. BIRCH TREE. The bark and Sap.—Upon deeply wounding or boring the trunk of the tree in the beginning of spring, a sweetish juice issues forth, sometimes, as is said, in so large quantity, as to equal in weigth to the whole tree and root: one branch will bleed a gallon or more a day. This juice is chiefly recommended in scorbutic disorders, and other foulnesses of the blood: its most sensible effect is to promote the urinary discharge.

317. BORAGO officinalis. BORAGE. The Flowers.—An exhilarating virtue has been attributed to the flowers of borage, which are hence ranked among the so called cordial flowers: but they appear to have very little claim to any virtue of this kind, and seem to be altogether insignificant.

318. BRYONIA alba. WHITE BRYONY. The Roots.—This is a strong irritating cathartic; and as such has sometimes been successfully exhibited in maniacal cases, in some kinds of dropsies, and in several chronical disorders, where a quick solution of viscid juices, and a sudden stimulus on the solids, were required.

319. CALENDULA officinalis. MARIGOLD. The Flowers.—These are supposed to be aperient and attenuating; as also cardiac, alexipharmic, and sudorific:

they are principally celebrated in uterine obstructions, the jaundice, and for throwing out the small-pox. Their sensible qualities give little foundation for these virtues: they have scarcely any taste, and no considerable smell. The leaves of the plant discover a viscid sweetishness, accompanied with a more durable saponaceous pungency and warmth: these seem capable of answering some useful purposes, as a stimulating, aperient, antiscorbutic medicine.

320. CANNABIS sativa. HEMP. The Seeds.—These have some smell of the herb; their taste is unctuous and sweetish; on expression they yield a considerable quantity of insipid oil: hence they are recommended (boiled in milk, or triturated with water into an emulsion) against coughs, heat of urine, and the like. They are also said to be useful in incontinence of urine; but experience does not warrant their having any virtues of this kind.

321. CARTHAMUS tinctorius. SAFFLOWER. The Seeds.—These have been celebrated as a cathartic: they operate very slowly, and for the most part disorder the bowels, especially when given in substance; triturated with aromatic distilled waters, they form an emulsion less offensive, yet inferior in efficacy to more common purgatives.

322. CENTAUREA Cyanus. BLUE-BOTTLE. The Flowers.—As to their virtues, notwithstanding the present practice expects not any from them, they have been formerly celebrated against the bites of poisonous animals, contagious diseases, palpitations of the heart, and many other distempers.

323. CENTAUREA rhapontica. GREATER CENTAURY. The Root.—It has a rough somewhat acrid taste, and abounds with a red viscid juice; its rough taste has gained it

some esteem as an astringent; its acrimony as an aperient; and its glutinous quality as a vulnerary: the present practice takes little notice of it in any intention.

324. CHELIDONIUM majus. GREAT CELANDINE. The Leaves and Juice.—This is an excellent medicine in the jaundice; it is also good against all obstructions of the viscera, and, if continued a time, will do great service against the scurvy. The juice also is used successfully for sore eyes, removing warts, &c. It should be used fresh, for it loses the greatest part of its virtue in drying.

325. CHENOPODIUM olidum. STINKING GOOSEFOOT. The Leaves.—Its smell has gained it the character of an excellent anti-hysteric; and this is the only use it is applied to. Tournefort recommends a spiritous tincture, others a decoction in water, and others a conserve of the leaves, as of wonderful efficacy in uterine disorders.

326. CHRYSANTHEMUM Leucanthemum. OX-EYE DAISY. The Leaves.—Geoffroy relates that the herb, gathered before the flowers have come forth, and boiled in water, imparts an acrid taste, penetrating and subtile like pepper; and that this decoction is an excellent vulnerary and diuretic.

327. CISTUS ladanifetus. GUM CISTUS.—The gum labdanum is procured from this shrub, and is its only produce used in medicine. This is an exudation from the leaves and twigs in the manner of manna, more than of any thing else. They get it off by drawing a parcel of leather thongs over the shrubs. It is not much used, but it is a good cephalic.—Hill's Herbal, p. 72.

328. CLEMATIS recta. UPRIGHT VIRGIN'S BOWER.—The whole plant is extremely acrid. It was useful for Dr.

Stoerck to employ the leaves and flowers in ulcers and cancers, as well as an extract prepared from the former; yet the preparation which he chiefly recommended was an infusion of two or three drams of the leaves in a pint of boiling water, of which he gave four ounces three times a-day, while the powdered leaves were applied as an escharotic to the ulcers.—Wood-ville's Med. Bot. p. 481.

329. COCHLEARIA Coronopus. SWINES-CRESS.—This is an excellent diuretic, safe and yet very powerful. The juice may be taken; and it is good for the jaundice, and against all inward obstructions, and against the scurvy: the leaves may also be eaten as sallet, or dried and given in decoction.—Hill's Hebal, p. 105.

330. CONVALLARIA Polygonatum. SOLOMON'S SEAL. The Root.—The root has several joints, with some flat circular depressions, supposed to resemble the stamp of a seal. It has a sweetish mucilaginous taste. As to its virtues, practitioners do not now expect any considerable ones from it, and pay very little regard to the vulnerary qualities which it was formerly celebrated for. It is used by pugilists to remove the black appearance occasioned from extravasated blood, and for curing bruises on the face, particularly black-eyes obtained by boxing.

331. CONVALLARIA majalis. MAY LILY. The Roots and Flowers.—The roots of this abound with a soft mucilage, and hence they have been used externally in emollient and maturating cataplasms: they were an ingredient in the suppurating cataplasm of the Edinburgh Pharmacopoeia. Those of the wild plant are very bitter: dried, they are said to prove a gentle errhine; as also are the flowers.

332. CONVOLVULUS sepium. BIND-WEED.—The poor people use the root of this plant fresh gathered and boiled in ale as a cathartic; and it is found generally to answer that purpose. It would, however, nauseate a delicate stomach; but for people of strong constitutions there is not a better medicine.

333. CUSCUTA europaea. DODDER. The whole plant gathered green is to be boiled in water with a little ginger and allspice, and this decoction operates as a cathartic; it also opens obstructions of the liver, and is good in the jaundice and many other disorders arising from the like cause.—Hill's Herbal.

334. CYNOGLOSSUM officinale. HOUNDS-TONGUE. The Root.—The virtues of this root are very doubtful: it is generally supposed to be narcotic, and by some to be virulently so: others declare that it has no virtue of this kind, and look upon it as a mere glutinous astringent.

335. CYPERUS longus. LONG CYPERUS. The Root.—This is long, slender, crooked, and full of knots: outwardly of a dark-brown or blackish colour, inwardly whitish; of an aromatic smell, and an agreeable warm taste: both the taste and smell are improved by moderate exsiccation. Cyperus is accounted a good stomachic and carminative, but is at present very little regarded.

336. DICTAMNUS albus. WHITE or BASTARD DITTANY. The Root.—The cortical part of the root, dried and rolled up into quills, is sometimes brought to us. This is of a white colour, a weak, not very agreeable smell; and a durable bitter, lightly pungent taste. It is recommended as an alexipharmic.

337. EQUISETUM palustre. HORSE-TAIL. The Herb.—It is said to be a very strong astringent: it has indeed a manifest astringency, but in a very low degree.

338. ERYSIMUM officinale.—It is said to be attenuant, expectorant, and diuretic; and has been strongly recommended in chronical coughs and hoarseness. Rondeletius informs us that the last-mentioned complaint, occasioned by loud speaking, was cured by this plant in three days. Other testimonies of its good effects in this disorder are recorded by writers on the Materia Medica, of whom we may mention Dr. Cullen; who for this purpose recommends the juice of the Erysimum to be mixed with an euqal quantity of honey and sugar; in this way also it is said to be an useful remedy in ulcerations of the mouth and throat.—Woodville's Med. Bot. p. 407.

339. ERYSIMUM Alliaria. SAUCE ALONE.—The leaves of this plant are very acrimonious, and have a strong flavour of onions. It is considered as a powerful diaphoretic, diuretic, and antiscorbutic.—Woodville's Med. Bot.

340. EUPATORIUM cannabinum. HEMP AGRIMONY, &c. Leaves.—They are greatly recommended for strengthening the tone of the viscera, and as an aperient; and said to have excellent effects in the dropsy, jaundice, cachexies, and scorbutic disorders. Boerhaave informs us, that this is the common medicine of the turf-diggers in Holland, against scurvies, foul ulcers, and swellings in the feet, which they are subject to. The roof of this plant is said to operate as a strong cathartic.

341. EUPHORBIA Esula. SPURGE FLAX. Its Berries.—These are useful in removing warts and excrescences, if

bruised and laid thereon. They are so acrid in their nature as to be altogether unfit for internal use.

342. EUPHRASIA officinalis. EYEBRIGHT. Leaves.—It was formerly celebrated as an ophtalmic, both taken internally and applied externally. Hildanus says he has known old men of seventy, who had lost their sight, recover it again by the use of this herb.

343. FRAGARIA vesca. THE STRAWBERRY. The Leaves and Fruit.—They are somewhat styptic, and bitterish; and hence my be of some service in debility and laxity of the viscera, and immoderate secretions, or a suppression of the natural evacuations depending thereon: they are recommended in haemorrhages and fluxes; and likewise as aperients, in suppressions of urine, obstructions of the viscera, in the jaundice, &c. The fruit is in general very grateful both to the palate and stomach: like other fruits of the dulco-acid kind, they abate heat, quench thirst, loosen the belly, and promote urine.

344. FUMARIA officinalis. FUMITORY. The Leaves.—The medical effects of this herb are, to strengthen the tone of the bowels, gently loosen the belly, and promote the urinary and other natural secretions. It is principally recommended in melancholic, scorbutic, and cutaneous disorders; for opening obstructions of the viscera, attenuating and promoting the evacuations of viscid juices.

345. GALEGA officinalis. GOAT'S RUE. The Herb.—This is celebrated as an alexipharmic; but its sensible qualities discover no foundation for any virtues of this kind: the taste is merely leguminous; and in Italy (where it grows wild) it is said to be used as food.

346. GALIUM Aparine. GOOSEGRASS, OR CLEAVERS. The Leaves.—It is recommended as an aperient, and in chronic eruptions; but practice has little regard to it.

347. GALIUM verum. LADIES BEDSTRAW, OR CHEESE-RENNET. The Herb.—This herb has a subacid taste, with a very faint, not disagreeable smell: the juice changes blue vegetable infusions to a red colour, and coagulates milk, thus exhibiting marks of acidity. It stands recommended as a mild styptic, and in epilepsy; but has never been much in use.

348. GERANIUM robertianum. HERB ROBERT. The leaves.—They have an austere taste, and have hence been recommended as astringent: but they have long been disregarded in practice.

349. GLECHOMA hederacea. GROUND-IVY. The Leaves.—This herb is an useful corroborant, aperient, and detergent; and hence stands recommended against laxity, debility, and obstructions of the viscera: some have had a great opinion of it for cleansing and healing ulcers of the internal parts, even of the lungs; and for purifying the blood. It is customary to infuse the dried leaves in malt liquors, to which it readily imparts its virtues; a practice not to be commended, unless it is for the purpose of medicine.

350. HEDERA helix. IVY. The Leaves and Berries.—The leaves have very rarely been given internally; notwithstanding they are recommended (in the Ephem. natur. curios. vol. ii. obs. 120.) against the atrophy of children; their taste is nauseous, acrid, and bitter. Externally they have sometimes been employed for drying and healing ichorous sores, and likewise for keeping issues open. The berries were supposed by the ancients to have a

purgative and emetic quality; later writers have recommended them in small doses, as diaphoretics and alexipharmics; and Mr. Boyle tells us, that in the London plague the powder of them was given with vinegar, with good success, as a sudorific. It is probable the virtue of the composition was rather owing to the vinegar than to the powder.

351. HERNIARIA glabra. RUPTUREWORT. The Leaves.—It is a very mild restringent, and may, in some degree, be serviceable in disorders proceeding from a weak flaccid state of the viscera: the virtue which it has been most celebrated for, it has little title to, that of curing hernias.

352. HYPERICUM perforatum. ST. JOHN'S WORT. The Leaves and Flowers.—Its taste is rough and bitterish; the smell disagreeable. Hypericum has long been celebrated as a corroborant, diuretic, and vulnerary; but more particularly in hysterical and maniacal disorders: it has been reckoned of such efficacy in these last, as to have thence received the name of fuga daemonum.

353. JASMINUM officinale. JASMINE. The Flowers.— The flowers have a strong smell, which is liked by most people, though to some disagreeable: expressed oils extract their fragrance by infusion; and water elevates somewhat of it in distillation, but scarcely any essential oil can be obtained from them: the distilled water, kept for a little time, loses its odour.

354. IRIS Pseudoacorus. FLOWER-DE-LUCE. The Root.—The roots, when recent, have a bitter, acrid, nauseous taste, and taken into the stomach prove strongly cathartic; and hence the juice is recommended in dropsies, in the dose of three or four scruples. By drying they lose

this quality, yet still retain a somewhat pungent, bitterish taste: their smell in this state is of the aromatic kind.

355. IRIS florentina. FLORENTINE IRIS, OR ORRIS-ROOT.—The roots grown in this country have neither the odour nor the other qualities that those possess which are grown in warmer climates: so that, for the purposes of medicine, they are usually imported from Leghorn.

The root in its recent state is extremely acrid, and, when chewed, excites a pungent heat in the mouth which continues several hours; but on being dried, this acrimony is almost wholly dissipated, the taste becomes slightly bitter, and the smell approaching to that of violets. It is now chiefly used in its dried state, and ranked as a pectoral or expectorant. The principal use of the roots is, however, for the purposes of perfumery, for which it is in considerable demand.

356. LACTUCA sativa. GARDEN LETTUCE. The Leaves and Seeds.—It smells strongly of opium, and resembles it in its effects; and its narcotic power, like that of the poppy heads, resides in its milky juice. An extract from the expressed juice is recommended in small doses in dropsy. In those diseases of long standing proceeding from visceral obstructions, it has been given to the extent of half an ounce a-day. It is said to agree with the stomach, to quench thirst, to be greatly laxative, powerfully diuretic, and somewhat diaphoretic.

357. LAMIUM album. WHITE ARCHANGEL, OR DEAD NETTLE. The Flowers.—The flowers have been particularly celebrated in female weaknesses, as also in disorders of the lungs; but they appear to be of very weak powers.

358. LAVENDULA Stoechas. ARABIAN STOECHAS, OR FRENCH LAVEN-DER. The Flowers.—They have a very fragrant smell, and a warm, aromatic, bitterish, subacrid taste: distilled with water, they yield a considerable quantity of a fragrant essential oil; to rectified spirit it imparts a strong tincture, which inspissated proves an elegant aromatic extract, but is seldom used in medicine.

359. LEONURUS Cardiaca. MOTHERWORT. The Leaves.—These have a bitter taste, and a pretty strong smell: they are supposed to be useful in hysteric disorders, to strengthen the stomach, to promote urine; and indeed it may be judged from their smell and taste, that their medical virtues are considerable, though they are now rejected both from the London and Edinburgh Pharmacopoeias.

360. LILIUM candidum. WHITE LILY. The Roots.—These are used in poultices. The good housewife doctors cut the roots in slices and steep them in brandy; and they are said to be an excellent remedy for all bruises and green wounds: for which purposes it is applied by them with considerable effect.

361. LITHOSPERMUM officinale. GROMWELL. The Seeds.—These are roundish, hard, and of a whitish colour, like little pearls. Powdered, they have been supposed peculiarly serviceable in calculous disorders. Their taste is merely farinaceous.

362. LYSIMACHIA Nummularia. MONEYWORT, OR HERB TWOPENCE. The Leaves.— Their taste is subastringent, and very slightly acid: hence they stand recommended by Boerhaave in the hot scurvy, and in uterine and other haemorrhagies. But their effects are so inconsiderable, that common practice takes no notice of them.

363. MALVA alcea. VERVAIN-MALLOW. The Leaves.—Alcea agrees in quality with the Althaea and Malva vulgaris; but appears to be less mucilaginous than either.

364. MATRICARIA Parthenium. COMMON WILD FEVERFEW. The Leaves and Flowers.—Simon Pauli relates, that he has experienced most happy effects from it in obstructions of the uterine evacuations. I have often seen, says he, from the use of a decoction of Matricaria and chamomile flowers with a little mugwort, hysteric complaints instantly relieved, and the patient from a lethargic state, returned as it were into life again. Matricaria is likewise recommended in sundry other disorders, as a warm stimulating bitter: all that bitters and carminatives can do, says Geoffroy, may be expected from this. It is undoubtedly a medicine of some use in these cases, though not perhaps equal to chamomile flowers alone, with which the Matricaria agrees in sensible qualities, except in being weaker.

365. NEPETA Calamintha. FIELD CALAMINT. The Leaves.—This is a low plant, growing wild about hedges and highways, and in dry sandy soils. The leaves have a quick warm taste, and smell strongly of pennyroyal: as medicines, they differ little otherwise from spearmint, than in being somewhat hotter, and of a less pleasant odour; which last circumstance has procured calamint the preference in hysteric cases.

366. NEPETA cataria. NEP, OR CATMINT. The Leaves.—This is a moderately aromatic plant, of a strong smell, not ill resembling a mixture of mint and pennyroyal; it is also recommended in hysteric cases.

367. NIGELLA romana. FENNEL-FLOWER. The Seeds.—They have a strong, not unpleasant smell; and a subacrid, somewhat unctuous disagreeable taste. They stand recommended as aperient, diuretic, &c. but being suspected to have noxious qualities should be used with caution.

368. NYMPHAEA alba. WHITE WATER-LILY. The Root and Flowers.—These have a rough, bitterish, glutinous taste, (the flowers are the least rough,) and when fresh a disagreeable smell, which is in great measure lost by drying: they are recommended in alvine fluxes, gleets, and the like. The roots are supposed by some to be in an eminent degree narcotic.

369. OCYMUM Basilicum. BASIL. The Leaves.—These have a soft, somewhat warm taste; and when rubbed, a strong unpleasant smell, which by moderate drying becomes more agreeable. They are said to attenuate viscid phlegm, promote expectoration, and the uterine secretions.

370. OPHIOGLOSSUM vulgatum. ADDERS-TONGUE. The Leaf.—An ointment is made of the fresh leaves, and it is a good application to green wounds. It is a very antient application, although now discarded from the apothecary's shop.

371. PAEONIA corolloides. MALE PEONY. The Seeds.—These are strong, and worn round the neck to assist detention, and are probably as good as other celebrated anodyne beads which have been so long recommended for the same purpose.

372. PHELLANDRIUM aquaticum. WATER HEMLOCK.—The seeds of this plant, according to Dr. Lange, when taken in large doses, produce a remarkable

sensation of weight in the head, accompanied with giddiness, intoxication, &c. It may probably prove, however, an active medicine, especially in wounds and inveterate ulcers of different kinds, and even in cancers; also in phthisis pulmonalis, asthma, dyspepsia, intermittent fevers, &c. About two scruples of the seed, two or three times a-day, was the ordinary dose given. Medicines of this kind should be used with great caution.—Woodville's Med. Bot. p. 91, 92.

373. PIMPINELLA saxifraga. BURNET SAXIFRAGE. The Root, Leaves, and Seeds.—This root promises from its sensible qualities, to be a medicine of considerable utility, though little regarded in common pratice. Stahl, Hoffman, and other German physicians, are extremely fond of it, and recommend it as an excellent stomachic, resolvent, detergent, diuretic, diaphoretic, and alexipharmic.

374. PLANTAGO major. COMMON BROAD-LEAVED PLANTAIN.—The leaves are slightly astringent, and the seeds said to be so; and hence they stand recommended in haemorrhages, and other cases where medicines of this kind are proper. The leaves bruised a little, are the usual application of the common people to slight flesh wounds. The Edinburgh College used to direct an extract to be made from the leaves.

375. POTENTILLA anserina. SILVERWEED. The Leaves.—The sensible qualities of Anserina promise no great virtue of any kind, for to the taste it discovers only a slight roughness, from whence it was thought to be entitled to a place among the milder corroborants. As the astringency of Tormentil is confined chiefly to its root, it might be thought that the same circumstance would take place in this plant; but the root is found to have no other

than a pleasant sweetish taste, like that of parsnip, but not so strong.

376. POTENTILLA reptans. CINQUEFOIL, OR FIVE-LEAVED GRASS. Root.—The root is moderately astringent: and as such is sometimes given internally against diarrhoeas and other fluxes; and employed in gargarisms for strengthening the gums, &c. The cortical part of the root may be taken, in substance, to the quantity of a dram: the internal part is considerably weaker, and requires to be given in double the dose to produce the same effect. It is scarcely otherwise made use of than as an ingredient in Venice treacle.—Lewis's Mat. Med.

377. POPULUS niger. THE BLACK POPLAR. Its Buds.—The young buds or rudiments of the leaves, which appear in the beginning of spring, abound with a yellow, unctuous odorous juice. They have hitherto been employed chiefly in an ointment, which received its name from them; though they are certainly capable of being applied to other purposes: a tincture of them made in rectified spirit, yields upon being isnpissated, a fragrant resin superior to many of those brought from abroad.

378. PRIMULA officinalis. COWSLIP. The Flowers.—The flowers appear in April; they have a pleasant sweet smell, and a subacrid, bitterish, subastringent taste. An infusion of them, used as tea, is recommended as a mild corroborant in nervous complaints. A strong infusion of them, with a proper quantity of sugar, forms an agreeable syrup, which for a long time maintained a place in the shops. By boiling, even for a little time, their fine flavour is destroyed. A wine is also made of the flowers, which is given as an opiate.

379. PRUNELLA vulgaris. SELFHEAL. The Leaves.—It has an herbaceous roughish taste, and hence stands recommended in haemorrhages and alvine fluxes. It has been principally celebrated as a vulnerary, whence its name; and in gargarisms for aphthae and inflammations of the fauces.

380. PULMONARIA officinalis. SPOTTED LUNGWORT. The Leaves.—They stand recommended against ulcers of the lungs, phthisis, and other like disorders.—Lewis's Mat. Med.

381. RANUNCULUS Ficaria. PILEWORT. The Leaves and Root.—The roots consist of slender fibres, with some little tubercles among them. These, with the leaves, are considered of considerable eficacy in the cure of haemorrhoids; for which purpose, considerable quantities are sold at herb-shops in London.

382. RANUNCULUS Flammula. SMALL SPEARWORT.—It has been lately discovered that this plant possesses very active powers as an emetic, and it is supposed to be useful in some cases of vegetable poisons.

383. RHAMNUS Frangula. THE BLACK OR BERRY-BEARING ALDER. Its Bark.—The internal bark of the trunk or root of the tree, given to the quantity of a dram, purges violently, occasioning gripes, nausea, and vomiting. These may be in good measure prevented by the addition of aromatics; but we have plenty of safer and less precarious purgatives.

384. RHUS coriaria. ELM-LEAVED SUMACH.—Both the leaves and berries have been employed in medicine; but the former are more astringent and tonic, and have been

long in common use, though at present discarded from the Pharmacopoeias.

385. RIBES nigrum.—The juice of black currants boiled up with sugar to a jelly, is an excellent remedy against sore throats.

386. RUMEX Hydrolapathum. THE GREAT WATER DOCK.—The leaves of the docks gently loosen the belly, and have sometimes been made ingredients in decoctions for removing a costive habit. The roots, in conjunction with other medicines, are celebrated for the cure of scorbutic and cutaneous disorders, for which the following receipt is given by Lewis.

Six ounces of the roots of the water dock, with two of saffron; and of mace, cinnamon, gentian root, liquorice root, and black pepper, each three ounces, (or, where the pepper is improper, six ounces of liquorice,) are to be reduced into coarse powder, and put into a mixture of two gallons of wine, with half a gallon of strong vinegar, and the yolks of three egs; and the whole digested, with a moderate warmth, for three days, in a glazed vessel close stopped: from three to six ounces of this liquor are to be taken every morning on an empty stomach, for fourteen or twenty days, or longer.

387. SALVIA Sclarea. GARDEN CLARY. The Leaves and Seeds.—These have a warm, bitterish, pungent taste; and a strong, not very agreeable smell: the touch discovers in the leaves a large quantity of glutinous or resinous matter. They are principally recommended in female weaknesses, in hysteric disorders, and in flatulent colics.

388. SAMBUCUS Ebulus. DWARF ELDER, OR DANEWORT. The Root, Bark, and Leaves.—These have a

nauseous, sharp, bitter taste, and a kind of acrid ungrateful smell: they are all strong cathartics, and as such are recommended in dropsies, and other cases where medicines of that kind are indicated. The bark of the root is said to be strongest: the leaves the weakest. But they are all too churlish medicines for general use: they sometimes evacuate violently upwards, almost always nauseate the stomach, and occasion great uneasiness of the bowels. By boiling they become (like the other drastics) milder, and more safe in operation. Fernelius relates, that by long coction they entirely lose their purgative virtue. The berries of this plant are likewise purgative, but less virulent than the other parts. A rob prepared from them may be given to the quantity of an ounce, as a cathartic; and in smaller ones as an aperient and deobstruent in chronic disorders: in this last intention, it is said by Haller to be frequently used in Switzerland, in the dose of a dram.

389. SANICULA officinalis. SANICLE. The Leaves.—These have an herbaceous, roughish taste: they have long been celebrated for sanative virtues, both internally and externally; nevertheless their effects, in any intention, are not considerable enough to gain them a place in the present practice.

390. SAPONARIA officinalis. SOAPWORT. The Herb and Root.—The roots taste sweetish and somewhat pungent; and have a light smell like those of liquorice: digested in rectified spirit they yield a strong tincture, which loses nothing of its taste or flavour in being inspissated to the consistence of an extract. This elegant root has not come much into practice among us, though it promises, from its sensible qualities, to be a medicine of considerable utility: it is greatly esteemed by the German physicians as an aperient, corroborant, and sudorific; and

preferred by the College of Wirtemberg, by Stahl, Neumann, and others, to sarsaparilla.

391. SAXIFRAGA granulata.—Linnaeus describes the taste of this plant to be acrid and pungent, which we have not been able to discover. Neither the tubercles of this root, nor the leaves, manifest to the organs of taste any quality likely to be of medicinal use; and therefore, though this species of Saxifraga has been long employed as a popular remedy in nephritic and gravelly disorders, yet we do not find, either from its sensible qualities or from any published instances of its efficacy, that it deserves a place in the Materia Medica.—Woodville's Med. Bot. p. 551.

392. SCABIOSA succisa. DEVIL'S BIT. The Leaves and Roots.—These stand recommended as alexipharmics, but they have long given place to medicines of greater efficacy.

393. SCANDIX Cerefolium. Chervil. The Leaves.—Geoffroy assures us, that he has found it from experience to be of excellent service in dropsies: that in this disorder it promotes the discharge of urine when suppressed, renders it clear when feculent and turbid, and when high and fiery of a paler colour; that it acts midly without irritation, and tends rather to allay than excite inflammation. He goes so far as to say, that dropsies which do not yield to this medicine are scarce capable of being cured by any other. He directs the juice to be given in the dose of three or four ounces every fourth hour, and continued for some time, either alone, or in conjunction with nitre and syrup.

394. SEDUM Telephium. ORPINE. The Leaves.—This is a very thick-leaved juicy plant, not unlike the houseleeks. It has a mucilaginous roughish taste, and hence is recommended as emollient and astringent, but has never been much regarded in practice.

395. SEMPERVIVUM tectorum. GREATER HOUSE-LEEK. The Leaves.—These are principally applied in cases of erysipelatous and other hot eruptions of the skin, in which they are of immediate service in allaying the pain arising therefrom: great quantities are cultivated in Surrey, and brought to the London markets. It is remarkable of this plant, that its juice, when purified by filtration, appears of a dilute yellowish colour upon the admixture of an equal quantity of rectified spirit of wine; but forms a beautiful white, light coagulum, like the finer kinds of pomatum: this proves extremely volatile; for when freed from the aqueous phlegm, and exposed to the air, it altogether exhales in a very little time.

396. SENECIO Jacobaea. RAGWORT. The Leaves.—Their taste is roughish, bitter, pungent, and extremely unpleasant: they stand strongly recommended by Simon Pauli against dysenteries; but their forbidding taste has prevented its coming into practice.

397. SOLANUM nigrum. COMMON NIGHTSHADE. The Leaves and Berries.—In the year 1757, Mr. Gataker, surgeon to the Westminster Hospital, called the attention of the Faculty to this plant, by a publication recommending its internal use in old sores, srophulous and cancerous ulcers, cutaneous eruptions, and even dropsies; all of which were much relieved or completely cured of it.

398. SPIRAEA Ulmaria. MEADOW-SWEET. The Leaves and Flowers.—The flowers have a very pleasant flavour, which water extracts from them by infusion, and elevates in distillation.

399. SPIRAEA Filipendula. DROPWORT. The Root.—The root consists of a number of tubercles, fastened together by slender strings; its taste is rough and bitterish,

with a slight degree of pungency. These qualities point out its use in a flaccid state of the vessels, and a sluggishness of the juices: the natural evacuations are in some measure restrained or promoted by it, where the excess or deficiency proceeds from this cause. Hence some have recommended it as an astringent in dysenteries, a diuretic, and others as an aperient and deobstruent in scrophulous habits.

400. SYMPHYTUM officinale. COMFREY. The Root.—The roots are very large, black on the outside, white within, full of a viscid glutinous juice, of no particular taste. They agree in quality with the roots of Althaea; with this difference, that the mucilage of it is somewhat stronger-bodied. Many ridiculous histories of the consolidating virtues of this plant are related by authors.

401. TAMUS communis. BLACK BRYONY.—The root is one of the best diuretics known in medicine. It is an excellent remedy in the gravel and all obstructions of urine, and other disorders of the like nature.

402. TANACETUM vulgare. TANSY. The Leaves.—These have a bitterish warm aromatic taste; and a very pleasant smell, approaching to that of mint or a mixture of mint and maudlin. Water elevates their flavour in distillation; and rectified spirit extracts it by infusion. They have been recommended in hysteric cases.

403. TEUCRIUM Chamaepitys. GROUND PINE. The Leaves.—These are recommended as aperient and vulnerary, as also in gouty and rheumatic pains.

404. THYMUS vulgaris. THYME. The Leaves and Flowers.—A tea made of the fresh tops of thyme is good in asthmas and diseases of the lungs. It is recommended against nervous complaints; but for this purpose the wild

thyme is preferable. There is an oil made from thyme that cures the tooth-ache, a drop or two of it being put upon lint and applied to the tooth; this is commonly called oil of origanum.

405. TRIGONELLA Foenum-graecum. FOENUGREEK. The Seeds.—They are of a yellow colour, a rhomboidal figure; have a disagreeable strong smell, and a mucilaginous taste. Their principal use is in cataplasms, fomentations, and the like, and in emollient glysters.

406. VERBASCUM Thapsus. MULLEIN. The Leaves and Flowers.—Their taste discovers a glutinous quality; and hence they stand recommended as an emollient, and is in some places held in great esteem in consumptions. The flowers of mullein have an agreeable, honeylike sweetness: an extract prepared from them by rectified spirit of wine tastes extremely pleasant.

407. VERBENA officinalis. COMMON WILD VERVAIN. The Leaves and Root.— This is one of the medicines which we owe to the superstition of former ages; the virtue it has been celebrated for is as an amulet, on which a pamphlet was some years ago published. It was recommended to wear the root by a ribband tied round the neck for the cure of the scrophula, and for which purpose, even now, much of the root is sold in London. As the age of superstition is passing by, it will be needless to say more on the subject at present.

408. VERONICA officinalis. MALE SPEEDWELL. The Leaves.—Hoffman and Joh. Francus have written express treatises on this plant, recommending infusions of it, drunk in the form of tea, as very salubrious in many disorders, particularly those of the breast.

Observations on the Drying and Preserving of Herbs, &c. for Medicinal Purposes.

The student who has paid attention to the subject described in the foregoing sections, will be struck with the admirable contrivance of Divine Wisdom; that has caused such astringent substances as are contained in the oak and Peruvian bark, to be produced from the same soil, and in a similar way to those mucilaginous and laxative ones which we find in the juice of the marsh-mallow, and the olive oil. It is not intended in this small elementary work to enter into any investigation of the primitive parts of the vegetable creation, or how such different particles are secreted. It may therefore suffice, that, although the science of vegetable physiology admits of many very beautiful and instructing illustrations, yet they only go so far as to prove to us, that the first and grand principle of vegetable life and existence, as well as of the formation of all organic substances, consists in a system of attraction and combination of the different particles of nature, as they exist and are imbibed from the soil and the surrounding atmosphere. Thus, during their existence, we observe a continual series of aggregation of substance; but no sooner does the principle of life become extinct, than the agents of decomposition are at work, dividing and selecting each different substance, and carrying it back from whence it came:—"From dust thou comest, and to dust thou shalt return." This, therefore, seems to be the sum total of existence; the explanation of which, with all its interesting ramifications, is more fully explained by the learned professors in what is called the science of chemistry.

As plants of all descriptions, and their several parts, form a link of that chain by which the welfare of the universe is connected, the industry of mankind is excited to preserve them for the different purposes to which they are applicable, in the oeconomy of human existence, to whose use the greater part of the animal and vegetable creation appears to be subservient. As men, then, and rational beings, it becomes our duty so to manage those things, when necessary, as to counteract as much as possible the decomposition and corruption which are natural to all organized bodies when deprived of the living principle.

We find that some vegetables are used fresh, but the greater part are preserved in a dry state; in which, by proper management, they can be kept for a considerable time afterwards, both for our own use as well as for that of others who reside at a distance from the place of their production.

In the preparation of the parts of plants for medicinal purposes, we should always have in view the extreme volatility of many of those substances, and how necessary it therefore is, that the mode of preparation and drying should be done as quickly as possible, in order to counteract the effects of the air and light, which continue to dissipate, without intermission, these particles, during the whole time that any vegetable, either fresh or dried, is left to its influence.

If we consider the nature of hops, which I shall take as an example, as being prepared in this way on the largest scale, we shall find they consist of three different principles; namely, an aroma, combined with an agreeable bitter taste, and a yellow colour; all of which properties are, by the consumers and dealers therein, expected to exist in the article after drying.

The art of drying hops, therefore, has been a subject of speculation for many years; and although we find the kiln apparatus for preserving them differ in many places, from the various opinions of the projectors, yet they are all intended for the same mode of action, i. e. the producing of a proper degree of heat, which must be regulated according to the state of the atmosphere at the gathering season, and the consequent quantity of the watery extract that the hops contain at the time: thus it is usual to have two kilns of different temperatures at work at the same time. It should, however, be observed, that the principal art of drying hops is in doing it as quickly as possible, so as not to injure them in their colour. As soon as they are dried, it is considered necessary to put them up into close and thick bags.

It should be observed, that all vegetables contain at every period of their growth two distinct species of moiture: the one called by naturalists the common juice, which is the ascending sap, and is replete with watery particles: the other is termed the proper juice, which having passed up through the leaves, and being there concocted and deprived of the watery part, contains the principle on which various properties and virtues of the plant depend. We therefore find that the operations above described only go to this, that the watery particles in the common juice should be evaporated, as being a part necessary to be got rid of; and the proper juice being of a volatile nature, the less time the plants are exposed for that purpose, the less of this precious material will be lost: and as those parts are flying off continually from all dried vegetables, there should be one general rule made with regard to their peparation; for, if we instance mint, balm, pennyroyal, &c., the longer these are kept in the open air, the weaker are they found to be in their several parts.

From hence we may naturally infer, that the usual mode in which the generality of herbs are dried, is not so good for the purpose, as one would be if contrived on similar principles, as, during the length of time necessary for the purpose, a great deal of the principal parts of the plants must of course be evaporated and lost; for little else is regarded than to dry them so as to prevent putrefaction. Although the generality of herbs met with are prepared as above described, yet in such articles as Digitalis, Hyoscyamus, Conium, Toxicodendron, &c., where the quantity necessary for a dose is so small, and so much depends on its action, practitioners are often obliged to prepare it themselves. I shall therefore relate the following mode as the best adapted to that purpose. The Digitalis is prepared by collecting the leaves in the summer, and stripping them off from the foot-stalks; these should be then carefully exposed to a slow heat, and the watery extract slowly thrown off; in which they should not be exposed to any great degree of heat, which by its action will deprive them of their fine green colour. When this is effected, the whole may be put in contact with a heat that will enable the operator to reduce it to a fine powder. And in order to keep it with its virtues perfect, it will be necessary to deprive it as much as possible of the influence of air and light. Hence it is preserved in close glass bottles which are coated, and also placed in a dark part of the elaboratory. Now, it is necessary that all plants intended to be used in a dried state, should be prepared and protected in a similar manner; and although it may be considered as a superfluous trouble, so far as regards the more common kinds, particular attention should be paid to these, when a small quantity is a dose, and an over-dose a certain poison.

Other kinds of vegetables require a certain degree of fermentation, as Tobacco. The prinicpal art of preserving it consists in this operation being duly performed; for which purpose, as soon as the leaves of the herb are fit, the foot-stalks are broken, and the leaves left on, in order for the moisture in part to be evaporated. Afterwards these are gathered and tied in handfuls, and hung up in the shade to dry; and when sufficiently divested of moisture, the bundles are collected together and laid in large boxes or tubs, in which these are fermented, and afterwards taken out again and dried; when it is found fit to pack up for the market.

The properties of Stramonium, which has been so much recommended for curing asthma, consist pricipally in the aroma, which is only to be preserved in a similar manner: and I have found from experience, that if the leaves are separated from the plant in a manner similar to that of tobacco, and the rest of the plant, noth roots, stalks, and seed-vessels, be slit and sufficiently dried in the sun or in an oven, and the whole fermented together, a very different article is the produce than what it is when dried in the usual way, and left entirely to the chance influence of the atmosphere.

In the common operation of hay-making it may also be observed, that the continued turning it over and admitting its parts to the action of the sun and the air, is for the purpose of getting rid of the watery particles contained in it; and the quicker this is done, the better it is. And although this operation is so essentially necessary, yet care should be taken at the same time, that it be not made too dry, so as to prevent a due degree of fermentation being allowed to take place in the rick. And it may be observed that the best grasses, or other plants used for hay, if made too dry, so as to prevent the natural fermentation which their proper juices will excite, can never make either palatable or nutritive food for cattle. Neither can the same be effected if the article is used in too small quantities. It should be observed, that herbs of all kinds should be gathered for peserving when in full bloom; but when roots or barks are recommended, these should be collected in the autumn months. The principles laid down for preserving dried plants generally, will apply to these parts also.

SECTION IX.—PLANTS USED FOR CULINARY PURPOSES.

"Man's first great ruling passion is to eat."

In the following section I have confined myself principally to such as are in cultivation. There are many of our indigenous plants which, in times of scarcity, and in other cases of necessity, are used as food by the people in the neighbourhood where they grow. But of these I shall make a separate list.

409. ARTICHOKE. Cynara Scolymus.—We have several varieties of this plant in cultivation; but the most approved are the large green and the globe. They are propagated by taking off the young suckers from the old roots in May, and planting them in a piece of rich land. Artichokes have been raised from seed, but they are seldom perfected in this country.

410. ARTICHOKE, JERUSALEM. Helianthus tuberosus.—Is cultivated for the sake of its tubers, similar to the potatoe; but they are not generally esteemed.

411. ASPARAGUS. Asparagus officinalis.—A very delicious vegetable in the spring, and well known to all amateurs of gardening.

There is a variety called the Gravesend Asparagus, and another called the Battersea; but it is the richness of the soil and manure that makes the only difference.

412. BASIL, SWEET. Ocymum Basilicum.—A pot-herb of considerable use for culinary purposes. It is an annual; and the seeds should be sown in a hot-bed in March, and

transplanted into the open ground. It is usually dried as other pot-herbs.

413. BEANS. Vicia Faba.—The varieties of the garden-beans are as follow:—

The early Mazagan and Longpod are planted in November. These will usually be fit for use in June.

The Windsor.
The Toker.
The Sword Longpod.
The Green Toker.
The White-blossomed.

These are sown usually in succession from January to March, and afford a continuance of crop during the season.

414. BEANS, FRENCH OR KIDNEY. Phaseolus vulgaris.—The kidney beans are of two kinds; such as run up sticks and flower on the tops. Of this description we have in cultivation the following:—

The Scarlet Runner. The Dutch Runner.

Both these are much esteemed.

Of dwarf kinds we have many varieties. The pollen of these plants is very apt to become mixed; and, consequently, hybrid kinds differing in the colour of the seeds are often produced. The season for sowing these is from April till June.

The Black, or Negro Beans. The Blue Dwarf. The Early Yellow. The Black

Speckled. The Red Speckled. The Magpie. The Canterbury.

All these varieties are good and early beans. The white Canterbury is the kind most esteemed for pickling; the other sorts being all of them more or less discoloured: and this kind is the sort generally sold for such purpose in the London markets.

415. BEET, RED. Beta vulgaris v. rubra.—The roots of this variety are used both in soups and for early spring salads: it is cultivated by sowing the seeds in March; and the roots are usually kept all winter.

The white beet is only a variety of the other; and it is the tops that are usually eaten of this kind as a substitute for spinach. Its culture is the same as that of the red kind.

416. BORECOLE. Brassica Rapa.—Of borecole we have two varieties; the purple, and green. The former is in much esteem amongst the Germans, who make a number of excellent dishes from it in the winter.

The culture is the same as for winter cabbage of other kinds.

417. BRUSSELS SPROUTS. Brassica Rapa.—This is also a useful variety of the cabbage species, which is very productive, forming a large number of beautiful small close-headed cabbages on their high stalks in the winter season. The seeds are sown in March.

418. BURNET. Poterium Sanguisorba.—The young leaves of this plant are eaten with other tender herbs in the spring, and are considered a wholesome addition to mustard, cress, corn-salad, &c.

419. CABBAGE. Brassica oleracea.—The varieties of cabbage are numerous. The most esteemed are,

The Early York. The Early Sugar-loaf. The Early Battersea. The Early Russia.

They are all sown in August, and planted out for an early summer-crop, and are usually in season in May and June.

The Large Battersea. The Red Cabbage. The Green Savoy. The White Savoy.

These are usually sown in March, and planted for a winter crop.

The use and qualities of the cabbage are too well known to need any further description.

420. CAULIFLOWER. Brassica oleracea var.—The varieties are,

The Early. The Late.

The early cauliflower is sown in the first week in September, and usually sheltered under bell or hand glasses during the winter. By this means the crop is fit for table in the months of May and June.

The late sort is usually sown in the month of March, and planted out for a succession to the first crop.

421. CAPERS. Capparis spinosa.—This is the flower-pod before it opens of the above shrub, and is only kept as an ornamental plant here. I am induced to notice this plant, as I have known some things used in mistake for capers that are dangerous. I once saw an instance of this, in the seed-

vessels of the Euphorbia Lathyris (which is a poisonous plant) being pickled by an ignorant person.

422. CAPSICUM. Capsicum annuum.—Cayenne pepper is made from a small variety of this plant.

We have many varieties cultivated here in hot-beds; namely, yellow and red, of various shapes, as long, round, and heart-shaped. All these are very useful, either pickled by themselves, or mixed with any other substances, as love-apple, radish pods, &c. to which they impart a very fine warm flavour.

423. CARROT. Daucus Carota.—

The Orange Carrot.—For winter use.

The Early Horn ditto.—For summer use.—The former is usually sown in March; the latter being smaller, and more early, is commonly raised on hot-beds. The Early Horn Carrot may likewise be sown in August, and is good all winter.

424. CELERY. Apium graveolens.—Celery is now so generally known as to render a description of the plant useless; nor need it be told, that the stalks blanched are eaten raw, stewed, &c. It should be used with great caution, if grown in wet land, as it has been considered poisonous in such cases. The season of sowing celery is in April. We have a variety of this, which is red, and much esteemed.

425. CELERIAC. This is a variety of the Apium graveolens. It is hollow in the stem, and the roots are particularly large: although this is much used in Germany, it is not so much esteemed by us as the celery.

426. CHAMPIGNON. Agaricus pratensis.—This plant is equal in flavour to the mushroom when boiled or stewed: it is rather dry, and has little or no scent whatever.

427. CHARDOONS. Cynara Cardunculus.—The gardeners blanch the stalks as they do celery; and they are eaten raw with oil, pepper, and vinegar; or, if fancy directs, they are also either boiled or stewed.

428. CHERVIL. Scandix Cerefolium.—This plant is so much used by the French and Dutch, that there is scarcely a soup or salad but what chervil makes part of it: it is grateful to the taste. See article oenanthe crocata in the Poisonous Plants.

429. CIVES. Allium Schoenoprasum.—This is an excellent herb for salads in the spring: it is also useful for soups, &c. &c. It is perennial, and propagated by its roots, which readily part at any season.

430. CLARY. Salvia Sclarea.—The seeds are sown in autumn. It is biennial. The recent leaves dipped in milk, and then fried in butter, were formerly used as a dainty dish; but now it is mostly used as a pot-herb, and for making an useful beverage called Clary Wine, viz.—Put four pounds of sugar to five gallons of water, and the albumen of three eggs well beaten; boil these together for about sixteen minutes, then skim the liquor; and when it is cool, add of the leaves and blossoms two gallons, and also of yeast half a pint; and when this is completed, put it all together into a vessel and stir it two or three times a-day till it has done fermenting, and then stop it close for two months: afterwards draw it into a clean vessel, adding to it a quart of good brandy. In two months it will be fit to bottle.

431. COLEWORT. Brassica oleracea var.—This is a small variety of the common cabbage, which is sown in June, and planted out for autumn and winter use. These are often found to stand the severe frosts of our winter when the large sort of cabbages are killed; but its principal use with gardeners is, to have a crop that will occupy the land after the beans and pease are over, and perhaps Colewort is the most advantageous for such purposes.

432. CORN SALAD. Valeriana Locusta.—An annual, growing wild in Battersea fields, and many other parts of this kingdom.

It is usually sown in August, and stands the winter perfectly well; it is very similar to lettuce, and is a good substitute for it in the spring and winter seasons.

433. COSTMARY Tanacetum Balsamita.—Is used as a herb in salad. This is a perennial plant of easy culture.

434. CRESS. Lepidium sativum.—There are two varieties of cress, the curled and common. This is an ingredient with mustard in early salads.

435. CRESS, AMERICAN. Erysimum Barbarea.—This is cultivated for salads, and is much esteemed. It is increased by sowing the seeds in the spring. This is only good in the winter and spring seasons.

436. CUCUMBERS. Cucumis sativus.—Many sorts of cucumbers are cultivated by gardeners. The most esteemed are,

The Southgate Cucumber. The Long Prickly. The Long Turkey. The White Spined.

The early crop is usually sown in hot-beds in the spring, and is a crop on which most gardeners have always prided themselves, each on his best mode of management of this crop. They will also grow if sown in April, and planted out in the open ground.

The short prickly cucumber is grown for gerkins.

437. DILL. Anethum graveolens.—This is similar to fennel, and used in pickling. It is esteemed useful as a medicinal herb also; which see.

438. ENDIVE. Cichorium Endivia.—Of this we have three varieties in cultivation.

The Green Curled. The White Curled. The Batavian, or Broad-leaved.

These are sown usually in June and July, and planted out for use in the autumn and winter. Endive is well known as forming a principal part of our winter salads; for which purpose, it is usual with gardeners to blanch it, by tying the plants up together, and laying them in dry places.

439. ESCHALOT. Allium ascalonium.—This species of allium is very pungent: its scent is not unpleasant, but is very strong, and, in general, it is preferred to the onion for making soups and gravies. It is propagated by planting the bulbs in September and October: they are fit to take up in May and June, when they are dried and kept for use.

440. FENNEL. Anethum Foeniculum.—The use of this plant is so well knwon in the kitchen, as to render an account of it useless. It is propagated by sowing seeds in the spring.

441. GARLICK. Allium sativum.—This is used in the art of cookery in various ways, for soups, pickles, &c. It is cultivated by planting the small cloves or roots in the month of October. It is fit to pull up in spring; and the roots are dried for use.

442. GOURD. Cucurbita Melopepo.—The inhabitants of North America boil the squash or melon gourds when about the size of small oranges, and eat them with their meat. The pulp is used with sour apples to make pies. In scarcity it is a good substitute for fruit.

443. KOHLRABBI, or TURNEP-ROOTED CABBAGE. Brassica Rapa var.—We have two kinds of this in cultivation; but although these are both much eaten in Germany, they are not esteemed with us: in fact, we have so many varieties of the cabbage kind all the year round for culinary purposes, that nothing could much improve them. In countries further north than we are, this is probably an acquisition, as, from its hardiness, it is likely to stand the frost better than some of the more delicate varieties.

444. LEEKS. Allium Porrum.—There are two kinds of leeks: the Welsh and London.

Leeks are used principally in soups; they partake much of the nature of onions, but for this purpose are in general more esteemed. This plant has been so long cultivated in this country, that its native place is not known.

The seeds are sown in the spring, and it is in use all the winter.

445. LETTUCE. Lactuca sativa.—The varieties of lettuce are many. They are,

Green Coss. White do. Silesia do. Brown do. Egyptian do. Brown Dutch. White Cabbage. Imperial. Hammersmith Hardy. Tennis-ball.

These are sown every summer month. The brown and Egyptian coss are sown in August, and commonly stand the winter; and in the spring are fit for use.

446. LOVE-APPLE. Solanum Lycopersicum.—The Portuguese and Spaniards are so very fond of this fruit, that there is not a soup or gravy but what this makes an ingredient in; and it is deemed cooling and nutritive. It is also called Tomatas, or Tomatoes.

The green fruit makes a most excellent pickle with capsicums and other berries. It is annual, and raised in hot-bed, and planted out.

447. MARJORAM, WINTER. Origanum vulgare.—This is used as a sweet herb, and is a good appendage to the usual ingredients in stuffing, &c. It is a perennial plant, and propagated by planting out its roots in the spring of the year.

448. MARJORAM, SWEET. Origanum Marjorana.—This is also used for the same purpose as the last mentioned. It is an annual, and not of such easy culture as the last, requiring to be raised from seeds in an artificial heat. It is usually dried and kept for use.

449. MARYGOLD. Calendula officinalis.—An annual plant usually sown in the spring. The petals of the flowers are eaten in broths and soups, to which they impart a very pleasant flavour.

450. MUSHROOM. Agaricus campestris.—Is cultivated and well known at our tables for its fine taste and utility in sauces. These plants do not produce seeds that can be saved; they are therefore cultivated by collecting the spawn, which is found in old hot-beds and in meadow lands.

Various methods have been lately devised for raising mushrooms artificially: but none seem to be equal to those raised in beds, as is described in all our books of gardening. Raising this vegetable in close rooms by fire heat has been found to produce them with a bad flavour; and they are not considered so wholesome as those grown in the open air, or when that element is admitted at times freely to the beds.

451. MUSTARD, WHITE. Sinapis alba.—This is sown early in the spring; to be eaten as salad with cress and other things of the like nature; it is of easy culture. A salad of this kind may be readily raised on a piece of thick woollen-cloth, if the seeds are strewed thereon and kept damp; a convenient mode practised at sea on long voyages. Cress and rap may be raised in the same manner.

452. ONION. Allium oleraceum.—The kinds of onions in cultivation are,

The Deptford. The Reading. The White Spanish. The Portugal. The Globe, and The Silver skinned.

All these varieties are usually sown in the spring of the year, and are good either eaten in their young state, or after they are dried in the winter. The silver skinned kind is mostly in use for pickling. The globe and Deptford kinds are remarkable for keeping late in the spring. A portion of all the other sorts should be sown, as they are all very good, and some kinds will keep, when others will not.

453. ONION, WELSH. Allium fistulosum.—This is sown in August for the sake of the young plants, which are useful in winter salads, and are more hardy than the other cultivated sorts.

454. PARSLEY. Petroselium vulgare.—A well known potherb sown in the spring; and the plants, if not suffered to go to seed, will last two years. See aethusa Cynapium, in Poisonous Plants.

455. PARSNEP. Pastinaca sativa.—This is a well known esculent root, and is raised by sowing the seeds in the spring.

456. PEA. Pisum sativum.—This is a well known dainty at our tables during spring and summer. The varieties in cultivation are,

Turner's Early Frame. Early Charlton. Golden Hotspur. Double Dwarf.

These are usually sown in November and December, and will succeed each other in ripening in June, if the season is fine, and afford a crop all that month.

The Dwarf Marrow-fat. The Royal Dwarf. The Prussia Blue. The Spanish Dwarf.

These varieties are usually sown in gardens when it is not convenient to have them grow up sticks, being all of a dwarf kind.

The Tall Marrow-fat. The Green Marrow-fat. The Imperial Egg Pea. The Rose, or Crown Pea. The Spanish Morotto. Knight's Marrow Pea. The Grey Rouncival. The Sickle Pea.

This last variety has no skin in the pods. These are used as kidney beans, as also in the usual way. These varieties are of very large growth, and are only to be cultivated when there is considerable room, and must be supported on sticks placed in the ground for that purpose. The grey pea is usually eaten when in a dry state boiled. Hot grey peas used to be an article of common sale among our itinerant traders in London streets, but it has been dropped for some years. One or other of the different kinds of the larger varieties should be put into the ground every three weeks from March to the 1st week in June, and a crop is thereby insured constantly till the beginning of October.

It should be remarked, that peas, as well as all vegetable seeds, are liable to sport and become hybrid sorts; some of which are at times saved for separate culture, and are called, when found good, by particular names; so that every twenty or thirty years many of the kinds are changed. Thus Briant, in his Flora Diaetetica, enumerates fourteen varieties, a few only of which bear the same name as those now in the list of the London seedsmen.

457. POMPION. Cucurbita Pepo.—This is of the gourd species, and grows to a large size. It is not much in use with us: but in the south of Europe the inhabitants use the pulp with some acid fruits for pastry, and it is there very useful. It is also sometimes used in a similar manner here with apples. Almost all the gourd species are similar in taste and nutriments when used this way.

458. PURSLANE. Portulaca oleracea.—Two kinds of Purslane, the green and the golden, are cultivated. These are eaten with vinegar, &c. the same as other salad oils, and are a fine vegetable in warm weather. The seeds are usually sown in the spring.

459. RADISH. Raphanus sativus.—The varieties in cultivation are,

The Early Scarlet. The Early Purple Short-top. The Salmon Radish. The White Turnip Radish. The Red Turnip Radish. The Black Spanish.

The above are sown almost every month in the year, and when the weather is fine, every good garden may have a supply all the year of those useful and wholesome vegetables.

The black Spanish radish is a large rooted variety usually sown in August, and is eaten in the winter season.

The poor labouring man's fare, which is usually eaten under the hedge of the field of his employment, is often accompanied with a dried onion; and was this root more known than it generally is, it would yield him, at the expense of two-pence, with a little labour in his cottage garden, an equally pleasant and more useful sauce to his coarse but happy meals. I have observed many instances of this oeconomy amongst the labouring classes in my youth, but fear it is not quite so commonly made use of in the present day.

460. RADISH, HORSE. Cochlearia Armoracea.—The root of this vegetable is a usual accompaniment to the loyal and standard English dishes, the smoking baron and the roast surloin; with which it is most generally esteemed.

It should not be passed unnoticed here, that this very grateful and wholesome root is not at all times to be eaten with impunity. One or two instances of its deleterious effects have been witnessed by my much esteemed friend Dr. Taylor, the worthy Secretary at the Society of Arts, and which he has communicated to me. I shall insert his own words, particularly as it may be the means of preventing the botanical student from falling into the same error, after arriving with the usual good appetite, from his recreative task of herborizing excursions. "Some gentlemen having ordered a dinner at a tavern, of which scraped horse-radish was one; some persons in company took a small quantity, and, dipping it in salt, ate of it: these were soon seized with a suppression of urine, accompanied with inflammation of the kidneys, which shortly after proved fatal to one of the company. The Doctor was consulted; but not knowing exactly the cause of the complaint, of course was at a loss to apply a remedy in time. But another circumstance of the like nature having come under his notice, and being apprized of it, by a well applied corrective medicine he recovered the patient. It should, therefore, be made a general observation, under such circumstances, and those are not the most unpleasant we meet with in our researches, 'never to eat horse-radish on an empty stomach.'"

461. RAMPION. Campanula Rapunculus.—This plant is remarkable for its milky juice. In France, it is cultivated for its roots, which are boiled and eaten with salads; but in England it is little noticed, except by the French cooks, who use it as an ingredient in their soups and gravies. It is propagated by planting its roots in the spring.

462. RHAPONTIC RHUBARB. Rheum Rhaponticum.—The radical leaf-stalks of this plant being thick and juicy, and having an acid taste, are frequently used in the spring as a substitute for gooseberries before they are ripe, in making puddings, pies, tarts, &c. If they are peeled with care, they will bake and boil very well, and eat agreeably.

463. ROCAMBOLE. Allium sativum.—The rocambole is merely the bulbs on the top of the flower-stalk of the garlic,

it being a viviparous plant. The flavour of this being somewhat different, is used in the kitchen under the above name.

464. SAGE. Salvia officinalis.—Of this we have two varieties, green and red. The latter is considered the best for culinary purposes: it is the well-known sauce for geese and other water-fowl. It is propagated by cuttings in the spring.

465. SALSAFY. Tragopogon porrifolium.—A biennial, sown in March, and is usually in season during winter. The roots are the parts used, which are very sweet, and contain a large quantity of milky juice: it is a good vegetable plain boiled, and the professors of cookery make many fine dishes of it.

466. SAVORY, SUMMER. Satureja hortensis.

467. SAVORY, WINTER. Satureja montana.

Both sorts are used for the same purposes, as condiments among other herbs for stuffing, and are well known to cooks. The former is an annual, and raised by sowing the seeds in March and April. The other, being perennial, is propagated either by the same means or by cuttings in the spring of the year. It is also dried for winter use.

468. SAVOY CABBAGE. Brassica oleracea, (var.)

The Green Savoy. The White or Yellow Savoy.

A well-known species of cabbage grown for winter use, and is one of our best vegetables of that season. It is raised by sowing the seeds in May, and planting the plants in any spot of ground in July after a crop of peas or beans. Savoys stand the frost better than most other kinds of cabbages with close heads.

469. SCORZONERA. Scorzonera tingitana.—The roots of this are very similar to salsafy, and its culture and use nearly the same.

470. SEA KALE. Crambe maritima.—This grows wild on our sea-coasts, particularly in Devonshire, where it has long been gathered and eaten by the inhabitants thereabouts. It was used also to be cultivated; but was in general lost to our gardens, till my late partner, Mr. Curtis, having paid a visit to his friend Dr. Wavell at Barnstaple, found it at that gentleman's table; and on his return he collected some seeds, and planted a considerable spot of ground with it at Brompton in 1792; at which time it was again introduced to Covent-Garden, but with so little successs, that no person was found to purchase it, and consequently the crop was useless.

This celebrated botanist, however, published a small tract on its uses and culture, which met with a considerable sale, and introduced it again to general cultivation.

The seeds should be sown in March, and the following year the plants are fit for forming plantations, when they should be put out in rows about three feet apart, and one foot in the row. The vegetable is blanched either by placing over the crowns of the root an empty garden-pot, or by earthing it up as is usually done with celery. It is easily forced, by placing hot dung on the pots; and is brought forward in January, and from thence till May.

It has been noticed of sea-kale, that, on eating it, it does not impart to the urine that strong and unpleasant scent which asparagus and other vegetables do.

471. SKIRRETS. Sium Sisarum.—The roots of this plant are very similar to parsneps, both in flavour and quality; they are rather sweeter, and not quite so agreeable to some palates. It is a biennial sown in March, and used all the winter.

472. SORREL, COMMON. Rumex Acetosa.—Bryant says the Irish, who are particularly fond of acids, eat the leaves with their milk and fish; and the Laplanders use the juice of them as rennet to their milk. The Greenlanders cure themselves of the scurvy, with the juice mixed with that of the scurvy-grass. The seeds may be sown, or the roots planted, in spring or autumn; it is not in general cultivation, but is to be found abundantly wild in meadows, &c.

473. SORREL, ROUND-LEAVED, or FRENCH. Rumex scutatus.—The leaves of the plant have more acidity in them than the common; and although not in general use, it is one of the best salad-herbs in the early part of the year: it is propagated in the same mode as the common sort.

474. SPINACH, Spinacia oleracea.—-Two sorts of this vegetable are cultivated. The Round-leaved, which is very quick in its growth, is sown for summer use; and if the seeds are put into the ground every three weeks, a constant succession is obtained while the weather is warm; but frost will soon destroy it.

The Prickly Spinach is not so quick in growth, and is hardy enough to stand our winters: it is therefore sown in August, and succeeds the round-leaved sort; and is a good vegetable all our winter months.

475. TARRAGON. Artemisia Dracunculus.—The leaves of this make a good ingredient with salad in the spring; and it also makes an excellent pickle. It is propagated by planting the small roots in spring or autumn, being a perennial.

476. THYME. Thymus vulgaris.—This is a well-known potherb used in broths and various modes of cookery: it is propagated by seeds and cuttings early in the spring.

477. TRUFFLES. Lycoperdon Tuber.—Not in cultivation. The poor people in this country find it worth their while to train up dogs for the purpose of finding them, which, by having some frequently laid in their way, become so used to it, that they will scrape them up in the woods; hence they are called Truffle-dogs. The French cooks use them in soups, &c. in the same manner as mushrooms. The truffle is mostly found in beech woods: I have mentioned this, because it is very generally met with at table, although it is not in cultivation.

478. TURNEPS. Brassica Rapa.—The varieties in use for garden culture are, the Early Dutch, the Early Stone, and the Mouse-tail Turnep. The culture and uses of the turnep are too well known to require any description.

The country people cut a raw turnep in thin slices, and a lemon in the same manner: and by placing the slices alternately with sugar-candy between each, the juice of the turnep is extracted, and is used as a pleasant and good remedy in obstinate coughs, and will be found to relieve persons thus afflicted, if taken immediately after each fit. Although this is one of the remedies my young medical friends may be led to despise, yet I would, nevertheless, advise them to make use of it when need occasions.

The yellow turnep is also much esteemed as a vegetable; but is dry, and very different in taste from any of the common kinds.

SECTION X.—CULINARY PLANTS NOT IN CULTIVATION.

The following section cannot be too closely studied by people in all ranks of life. Many of our most delicate vegetables are found growing wild; and in times of scarcity, and after hard winters, many articles of this department will be found highly acceptable to all, and the condition of the poorer classes would be bettered by a more intimate knowledge of those plants. In fact, these and the medicinal plants ought to be known to every one: and in order to facilitate the study of them, I have been thus particular in my description of the different kinds.

479. AGARIC, ORANGE. Agaricus deliciosus.—This agaric well boiled and seasoned with pepper and salt, has a flavour similar to that of a roasted muscle. In this way the French, in general, make use of it. It is in high perfection about September, and is chiefly to be found in dry woods.

480. ALEXANDERS. Smyrnium Olustratum.—If the poorer people were aware of the value of this plant, which is now quite neglected, it might be turned to good account as an article of food, and that, in all likelihood, of the most wholesome kind.

Bryant thinks it was much esteemed by the monks, and states that it has, ever since the destruction of the abbeys in this country, remained in many places growing among the rubbish; hence the reason of its being found wild in such places.

481. ALEXANDERS, ROUND-LEAVED. Smyrnium perfoliatum.—-It is said that the leaves and stalks boiled

are more pleasant to the taste than the other kind of Alexanders.

482. ARROWHEAD. Sagittaria sagittifolia.—The roots of this plant are said to be very similar to the West-India arrow-root. They are sometimes dried and pounded, but are reported to have an acrid unpleasant taste; but this might perhaps be got rid of by washing the powder in water.

483. BLACKBERRY. Rubus fruticosus.—The berries of this plant are well known in the country; but if too many be eaten, they are apt to cause swelling in the stomach, sickness, &c.

484. BRIONY, BLACK. Tamus communis.—Although this is considered a poisonous plant, the young leaves and shoots are eaten boiled by the common people in the spring.

485. BURDOCK. Arctium Lappa.—Mr. Bryant in his Flora Diaetetica says that many people eat the tenders talks of this plant boiled as asparagus.

486. BURNET. Sanguisorba officinalis.—The young leaves form a good ingredient in salads. They have somewhat the flavour of cucumbers.

487. BUTTERWORT. Pinguicula vulgaris.—The inhabitants of Lapland and the north of Sweden give to milk the consistence of cream by pouring it warm from the cow upon the leaves of this plant, and then instantly straining it and laying it aside for two or three days till it acquires a degree of acidity.

This milk they are extremely fond of; and once made, they need not repeat the use of the leaves as above, for a spoonful or less of it will turn another quantity of warm milk, and make it like the first, and so on, as often as they please to renew their food.—Lightfoot's Flor. Scot. p. 77.

488. CHAMPIGNON. Agaricus pratensis.—There is little or no smell to be perceived in this plant, and it is rather dry; yet when boiled or stewed it communicates a good flavour, and is equal to the common mushroom.

489. CHANTARELLE. Agaricus Chantarellus.—This agaric, when broiled with pepper and salt, has a taste very similar to that of a roasted cockle, and is considered by the French a great delicacy. It is found principally in woods and old pastures, and is in good perfection about the middle of September.

490. CHARLOCK. Sinapis arvensis.—The young plant is eaten in the spring as turnep-tops, and is considered not inferior to that vegetable. The seeds of this have sometimes been saved and sold for feeding birds instead of rape; but being hot in its nature, it has been known to cause them to be diseased.

491. CHICKWEED. Alsine media.—This is a remarkably good herb boiled in the spring; a circumstance not sufficiently attended to.

492. CLOUD-BERRY. Rubus Chamaemorus.—This plant grows wild in some parts of the north of England: the fruit has nearly the shape of the currant, and is reckoned in Norway, where it grows abundantly, a favourite dish.

493. COTTON-THISTLE. Onopordon Acanthium.—The tender stalks of this plant, peeled and boiled, are by some considered good; but it has a peculiar taste which is not agreeable to all.

Bryant in his Flora Diaetetica says that the bottoms of the flowers are eaten as artichokes.

494. COW-PARSNEP. Heracleum Sphondylium.—The inhabitants of Kamschatka about the beginning of July collect the foot-stalks of the radical leaves of this plant, and, after peeling off the rind, dry them separately in the sun; and then tying them in bundles, they lay them up carefully in the shade. In a short time afterwards, these dried stalks are covered over with a yellow saccharine efflorescence tasting like liquorice, and in this state they are eaten as a delicacy.

The Russians, not content with eating the stalks thus prepared, contrive to get a very intoxicating spirit from them, by first fermenting them in water with the greater bilberry (Vaccinium uliginosum), and then distilling the liquor to what degree of strength they please; which Gmelin says is more agreeable to the taste than spirits made from corn. This may, therefore, prove a good succedaneum for whisky, and prevent the consumption of much barley, which ought to be applied to better purposes. Swine and rabbits are very fond of this plant.—-Lightfoot's Fl. Scot.

495. DANDELION. Leontodum Taraxacum.—This is a good salad when blanched in the spring. The French, who eat more vegetables than our country people do, use this in the spring as a common dish: it is similar to endive in taste.

496. DEWBERRY. Rubus caesius.—The dewberry is very apt to be mistaken for the blackberry; but it may be easily distinguished by its fruit being not so large, and being covered with blue bloom similar to that seen on plums: it has a very pleasant taste, and is said to communicate a grateful flavour to red wine when steeped in it.

497. EARTH-NUT. Bunium Bulbocastanum.—The roots are eaten raw, and considered a delicacy here, but thought much more of in Sweden, where they are an article of trade: they are eaten also stewed as chesnuts.

498. ELDER. Sambucus nigra.—The young shoots of elder are boiled with other herbs in the spring and eaten; they are also very good pickled in vinegar. Lightfoot says, in some countries they dye cloth of a brown colour with them.

499. FAT-HEN. Chenopodium viride et album.—These are boiled and eaten as spinach, and are by no means inferior to that vegetable.

500. FUCUS, SWEET. Fucus saccharatus.—This grows upon rocks and stones by the sea-shore. It consists of a long single leaf, having a short roundish foot-stalk, the leaf representing a belt or girdle. This is collected and eaten the same as laver, as are also the two following kinds.

501. FUCUS, PALMATED. Fucus palmatus.—This plant also grows by the sea-side, and has a lobed leaf.

502. FUCUS, FINGERED. Fucus digitatus.—This is also to be found by the sea-side, growing upon rocks and stones; it has long leaves springing in form of fingers when spread.

503. GOOD KING HENRY. Chenopodium Bonus-Henricus.—The leaves and stalk of this plant are much esteemed. The plant was used to be cultivated, but of late years it has been superseded by the great number of other esculent vegetables more productive than this. The young shoots blanched were accounted equal to asparagus, and were made use of in a similar manner.

504. HEATH. Erica vulgaris.—Formerly the young tops are said to have been used alone to brew a kind of ale; and even now, I am informed, the inhabitants of Isla and Jura (two islands on the coast of Scotland) continue to brew a very potable liquor, by mixing two-thirds of the tops of heath with one of malt.—Lightfoot's Fl. Scot.

505. HOPS. Humulus Lupulus.—Independently of the great use of hops in making beer, and for medicinal uses, where the plant grows wild, it affords the neighbours a dainty in the spring months. The young shoots, called hop-tops, when boiled, are equal in flavour to asparagus, and are eagerly sought after for that purpose.

506. LADIES-SMOCK. Cardamine pratensis.—This is good as a salad herb.

507. LAVER. Fucus esculentus.—This is collected by sailors and people along the sea-coasts; is eaten both raw and boiled, and esteemed and excellent antiscorbutic. The leaves of this Fucus are very sweet, and, when washed and hanged up to dry, will exude a substance like that of sugar.

508. MAPLE. Acer Pseudo-platanus.—By tapping this tree it yields a liquor not unlike that of the birch-tree, from which the Americans make a sugar, and the Highlanders sometimes an agreeable and wholesome wine.—Lightfoot's Fl. Scot.

509. MARSH MARIGOLD. Caltha palustris.—The flower-buds, before opening, are picked, and are considered a good substitute for capers.

510. MEADOW-SWEET. Spiraea Filipendula.—The roots of this, in Sweden, are ground and made into bread.

511. MILK-THISTLE. Carduus marianus.—The young leaves in the spring, cut close to the root with part of the stalks on, are said to be good boiled.

512. MOREL. Phallus esculentus.—The morel grows in wet banks and moist pastures. It is used by the French cooks, the same as the truffle, for gravies, but has not so good a flavour: it is in perfection in May and June.

513. MUSHROOM, VIOLET. Agaricus violaceus.—This mushroom requires more broiling than all the rest; but when well done and seasoned, it is very good. It is found in dry woods, old pastures, &c. where it grows to a large size.

514. MUSHROOM, BROWN. Agaricus cinnamomeus.—The whole of this plant has a nice smell, and when stewed or broiled has a pleasant flavour. It is to be found as the one above, and is fit for use in October.

515. ORPINE. Sedum telephium.—The leaves are eaten in salads, and are considered equal to purslane.

516. OX-TONGUE, COMMON. Picris Echioides.—The leaves are said to be good boiled.

517. PEAS, EARTH-NUT. Orobus tuberosus.—The roots of this, when boiled, are said to be nutritious. The Scotch Highlander chews the root as a substitute for tobacco.

518. PILEWORT. Ranunculus Ficaria.—The young leaves in spring are boiled by the common people in Sweden, and eaten as greens. The roots are sometimes washed bare by the rains, so that the tubercles appear above ground; and in this state have induced the ignorant in superstitious times to fancy that it has rained wheat, which these tubercles sometimes resemble.

519. SALEP. Orchis Morio.—The powder of these roots is used for a beverage of that name. This is imported chiefly from Turkey. It grows in this country, although it is never noticed: the roots are smaller than those imported, but will answer the purpose equally well.

520. SALTWORT. Salicornia europaea.—This is gathered on the banks of the Thames and Medway, and brought to London, where it is sold as samphire. It makes a very good pickle, but by no means equal to the true kind.

521. SAMPHIRE. Crithmum maritimum.—This has long been in much esteem as a pickle: it grows on the high cliffs on the Kentish coast, where people make a trade of collecting it by being let down from the upper part in baskets. A profession of great danger.

522. SCURVY-GRASS. Cochlearia officinalis.—The leaves are hot and pungent, but are considered very good, and frequently eaten between bread and butter.

523. SAUCE ALONE. Erysimum Alliaria.—This is very good boiled with salt-meat in the spring, when other vegetables are scarce. It is valuable to the poor people; and is, in general, a common plant under hedges.

524. SEA BINDWEED. Convolvulus Soldanella.—This plant is to be found plentifully on our maritime coasts, where the inhabitants plucks the tender stalks, and pickle them. It is considered to have a cathartic quality.

525. SEA-PEAS. Pisum maritimum.—These peas have a bitterish disagreeable taste, and are therefore rejected when more pleasant food is to be got. In the year 1555 there was a great famine in England, when the seeds of this plant

were used as food, and by which thousands of families were preserved.

526. SEA-WORMWOOD. Artemisia maritima.—Those who travel the country in searching after and gathering plants, if they chance to meet with sour or ill-tasted ale, may amend it by putting an infusion of sea-wormwood into it, whereby it will be more agreeable to the palate, and less hurtful to the stomach.—Threlkeld. Syn. Pl. Hibern.

This is an ingredient in the common purl, the usual morning beverage of our hardy labouring men in London.

527. SEA-ORACH, GRASS-LEAVED. Atriplex littoralis.—This plant is eaten in the same manner as the Chenopodium.

528. SEA-BEET. Beta maritima.—This is a common plant on some of our sea-coasts. The leaves are very good boiled, as are also the roots.

529. SILVER-WEED. Potentilla anserina.—The roots of this plant taste like parsneps, and are frequently eaten in Scotland either roasted or boiled.

In the islands of Tiras and Col they are much esteemed, as answering in some measure the purposes of bread, they having been known to support the inhabitants for months together during a scarcity of other provisions. They put a yoke on their ploughs, and often tear up their pasture-grounds with a view to get the roots for their use; and as they abound most in barren and impoverished soils, and in seasons when other crops fail, they afford a most seasonable relief to the inhabitants in times of the greatest scarcity. A singular instance this of the bounty of Providence to these islands.—Lightfoot's Fl. Scot.

530. SOLOMON'S-SEAL. Convallaria Polygonatum.—The roots are made into bread, and the young shoots are eaten boiled.

531. SPATLING-POPPY. Cucubalus Behen.—Our kitchen-gardens scarcely afford a better-flavoured vegetable than the young tender shoots of this when boiled. They ought to be gathered when they are not above two inches long. If the plant was in cultivation, no doubt but what it would be improved, and would well reward the gardener's trouble: it sends forth a vast quantity of sprouts, which might be nipped off when of a proper size; and there would be a succession of fresh ones for at least two months.

It being a perennial too, the roots might be transplanted into beds like those of asparagus.—Bryant's Fl. Diaetetica, p. 64.

532. SPEEDWELL. Veronica spicata.—This is used by our common people as a substitute for tea, and is said to possess a somewhat astringent taste, like green tea.

533. SPOTTED HAWKWEED. Hypochaeris maculata.—The leaves are eaten as salad, and are also boiled.

534. STINGING-NETTLE. Urtica dioica.—The young shoots in the spring are eaten boiled with fat meat, and are esteemed both wholesome and nutritive.

535. SHRUBBY STRAWBERRY. Rubus arcticus.—The fruit of this plant is very similar in appearance to a strawberry: its odour is of the most grateful kind; and its flavour has that delicate mixture of acid and sweet, which is not to be equalled by our best varieties of that fruit.

536. SWEET CICELY. Scandix odorata.—The leaves used to be employed in the kitchen as those of cervil. The green seeds ground small, and used with lettuce or other cold salads, give them an agreeable taste. It also grows in abundance in some parts of Italy, where it is considered as a very useful vegetable.

537. WATER-CRESS. Sisymbrium Nasturtium.—A well known herb in common use, but is not in cultivation, although it is one of our best salads.

538. WILLOW-HERB. Epilobium angustifolium.—The young shoots of these are eaten as asparagus.

SECTION XI.—PLANTS USEFUL IN DYEING.

There is no department of the oeconomy of vegetables in which we are more at a loss than in the knowledge of their colouring principles; and as this subject presents to the student an opportunity of making many interesting and useful experiments, I trust I shall stand excused, if I enter more fully into the nature of it than I have found it necessary to do in some of the former sections.

The following list of plants, which is given as containing colours of different kinds, are the same as have been so considered for many years past: for, latterly, little has been added to our stock of knowledge on this head. It may however be proper to observe, that a great number of vegetables still contain this principle in a superior degree, and only want the proper attention paid to the abstracting it.

Most of our dyeing drugs are from abroad; and even the culture of madder, which was once so much grown by our farmers, is now lost to us, to the great advantage of the Dutch, who supply our markets. But there is no reason why the agriculturist, or the artisan, should be so much beholden to a neighbouring nation, as to pay them enormous prices for articles which can be so readily raised at home; and, according to the general report of the consumers, managed in a way far superior to what it generally is when imported.

Let the botanical student therefore pay attention to this particular; for it is a wide field, in which great advantages may be reaped, either in this country or in any other part of the world where he may hereafter become an inhabitant.

The art of dyeing, generally considered, is kept so great a secret, that few persons have had the opportunity of making experiments. The extracting colours from their primitive basis is a chemical operation, and cannot be expected in this place; but as some persons may be inclined to ascertain these properties of vegetables, I shall go just so far into the subject as to give an idea of the modes generally used; and to state the principles on which the colouring property is fixed when applied to the purposes of dyeing cloth.

In the article Madder, page 32, I mentioned having made an extract similar to the Adrianople red. For which purpose, a sufficient quanitity of the roots should be taken fresh out of the ground, washed clean from the dirt, bruised in a mortar, and then boiled in rain-water till the whole becomes tinged of a red colour, then put into a cloth and all the colouring matter pressed out. This should again be put into hot water in a clean glazed earthen-pan, to which should be added a small quantity of water in which alum had been dissolved, and the whole stirred up together; then immediately add a lump of soda or pot-ash, stirring the whole up, when an effervescence will take place, the allum that had united with the juice of the madder will be found to become neutralized by the pot-ash, and the result will be a precipitate of the red fecula. This may be washed over in different waters, and either put by for use in a liquid state, or filtered and dried in powder or cakes. Most vegetable colours will not, however, admit of being extracted by water, and it is necessary to use an acid for that purpose: vinegar is the most common. But in making the extract from roots with acids, great care should be taken that they are sufficiently cleared from mould, sand, &c.; for, if the same should contain either iron, or any metallic substance, its union with the acid will cause a blackness, and of course spoil the tint. In a similar mode are all the different colouring principles extracted, either from leaves, flowers, fruits, or woods. The preparation of woad is a curious process on similar principles; which see in page 31.

Weld, or dyers weed, is generally used after it is dried. The whole plant is ground in a mill, and the extract made by boiling it. It is then managed with alum and acids agreeably to the foregoing rules, which are necessary for throwing out the colour.

Instructions how Substances may be tried, whether they are serviceable in Dyeing, from Hopson's Translation of Weigleb's Chemistry.

"In order to discover if any vegetable contains a colouring principle fit for dyeing, it should be bruised and boiled in water, and a bit of cotton, linen, or woollen stuff, which has previously been well cleaned, boiled in this decoction for a certain time, and rinsed out and dried. If the stuff becomes coloured, it is a sign that the colour may be easily extracted; but if little or no colour be perceived, we are not immediately to conclude that the body submitted to the trial has no colour at all, but must first try how it will turn out with the addition of saline substances. It ought, therefore, to be boiled with pot-ash, common salt, sal ammoniac, tartar, vinegar, alum, or vitriol, and then tried upon the stuff: if it then exhibit no colour, it may safely be pronounced to be unfit for dyeing with. But if it yields a dye or colour, the nature of this dye must then be more closely examined, which may be done in the following manner:—

Let a saturated decoction of the colouring substance be well clarified, distributed into different glass vessels, and its natural colour observed. Then to one portion of it let there be added a solution of common salt; to the second, some sal ammoniac; and to the third, alum; to the fourth, pot-ash; to the fifth, vitriolic or marine acid; and to the sixth, some green vitriol: and the mixtures be suffered to stand undisturbed for the space of twenty-four hours. Now in each of these mixtures the change of colour is to be observed, as likewise whether it yields a precipitate or not.

If the precipitate by the pure acid dissolve in an alkaline lixivium entirely, and with a colour, they may be considered as resino- mucilaginous particles, in which the tingeing property of the body must be looked for, which, in its natural state, subsists in an alkalino-saponaceous compound. But if the precipitate be only partly dissolved in this manner, the dissolved part will then be of the nature of a resinous mucilage, which in the operation has left the more earthy parts behind. But if nothing be precipitated by the acids, and the colour of the decoction is rendered brighter, it is a mark of an acido-mucilaginous compound, which cannot be separated by acids. In this there are mostly commonly more earthy parts, which are soon made to appear by the addition of an alkali.

When, in the instances in which green vitriol has been added, a black precipitate is produced, it indicates an astringent earthy compound, in which there are few mucilaginous particles. The more the colour verges to black, the more of this acid and mucilaginous substance will be found in it.

The mixture of alum with a tingeing decoction shows by the coloured precipitate that ensues from it, on the one hand, the colour it yields, and on the other hand, by the precipitate dissolving either partly or entirely in a strong alkaline lixivium, whether or not some of the earth of alum has been precipitated together with the colouring particles. Such substances as these must not, in general, be boiled with alum, although this latter ingredient may be very properly used in the preparation of the stuff.

When a tingeing decoction is precipitated by an alkaline lixivium, and the precipitate is not redissolved by any acid, for the most part neither one nor the other of these saline substances ought to be used, but the neutral salts will be greatly preferable. In all these observations that are made with respect to the precipitation effected by means of different saline substances, attention must be paid at the same time to the change of colour which ensues, in order to discover whether the colour brightens, or entirely changes.

When the colour of a decoction is darkened by the above-mentioned additions without becoming turbid, it shows that the colouring matter is more concentrated and inspissated. When the colour is brightened, a greater degree of solution and attenuation has taken place in the colouring matter in consequence of the addition. If the colour becomes clearer, and after a little time some of the tingeing substance is separated, it shows that part of the colour is developed, but that another part has been set loose from its combination by the saline substance.

But if the colouring matter is separated in great abundance by the saline addition, (the colour being brightened at the same time,) it may be considered as a sign that the colouring substance is entirely separated from the decoction, and that only an inconsiderable part, of a gummy nature, remains behind united with the additaments, which is in a very diluted state.—This is an effect of the solution of tin, as also sometimes of the pure acids.

If, indeed, a portion of the colouring substance be separated by a saline addition, but the rest of the colouring decoction becomes not-withstanding darker, it shows that the rest of the colouring particles have been more concentrated, and hence have acquired a greater power of tingeing. With regard to the proportion of the addition, the following circumstances may serve by way of guide:

When the colour of a decoction is darkened by the addition, without any precipitate being produced, no detriment can easily arise from using a redundancy of it, because the colour will not be further darkened by it. But if the colour be required to be brighter, the trial must first be made, which is the proportion by which the colour is darkened the most, and then less of it must be employed.

When the colour of a decoction is brightened by an addition without a precipitation ensuing, this addition can never be used in a larger quantity without hurting the colouring particles; because the colouring particles would be made too light, and almost entirely destroyed.—Such is the consequence of too large an addition of the solution of tin or of a pure acid.

When the addition produces a brighter colour, and part only of the colouring substance is separated without a further addition occasioning a fresh separation, somewhat more of it than what is wanted may be added to produce the requisite shading; because experience shows that, by this means, a greater quantity of tingeing particles is united with the woolly fibres of the cloth, and is capable of being, as it were, concentrated in them: for which purpose, however, these barks must be boiled down. This effect is chiefly observed with sal ammoniac and wine vinegar.

When by an addition which causes a separation of the colouring substance the colour becomes brighter in proportion the more there is used of it, it must be employed in a moderate quantity only; because otherwise, more and more of the colouring substance will be separated, and its tingeing power diminished. But when a colour is rendered dark at first by an addition, and afterwards, upon more of the same substance being added, becomes brighter, and this in proportion to the quantity that is added, it will be found that the darkening power has its determined limits; and that, for producing the requisite degree of darkness, neither too much nor too little must be taken.

To the before-mentioned principles also, the different proofs bear a reference, by which the fixity and durability of the colour with which a stuff has been dyed may be tried. Of these, some may be called natural, other artificial. The natural proof consists in exposing the dyed stuff to the air, sun, and rain. If the colour is not changed by this exposure in twelve or fourteen days, it may be considered as genuine; but if it is, the contrary is allowed. This proof, however, is not adapted to every colour; because some of them resist it, and yet will fade in consequence of the application of certain acids; others, on the contrary, that can not resist the natural proof remain unchanged by the latter. Colours, therefore, may be arranged in three classes; and to each of these a particular kind of artificial proof allotted. The first class is tried with alum, the second with soap, and the third with tartar.

For the proof with alum: Half an ounce of this is dissolved in one pound of boiling water in an earthenware vessel; into this is put, for instance, a drachm of yarn or worsted, or a piece of cloth of about two fingers breadth; this is suffered to boil for the space of five minutes, and is then washed in clean water. In this manner are tried crimson, scarlet, flesh-colour, violet, ponceau, peach-blossom colour, different shades of blue, and other colours bordring upon these.

For the proof with soap: Two drachms of this substance are boiled in a pint of water, and the small piece of dyed stuff that is to be tried is put into it, and likewise suffered to boil for the space of five minutes. With this all sorts of yellow, green, madder-red, cinnamon, and similar colours, are tried.

In the same manner is made the proof with tartar; only this should be previously pounded very small, in order that it may be more easily dissolved. With this all colours bordering upon the fawn are tried.

From the above we discover that the art of applying and fixing colours in dyeing depends on the chemical affinity between the cloth and the dyeing principle: and accordingly as this is more or less strong, so is the facility with which the substance is coloured, and on this the deepness of the dye depends: for frequently one kind of cloth will be found to receive no colour at all, whilst another will receive from the same composition a deep tinge. Cotton, for instance, receives scarcely any tinge from the same bath that will dye woollen a deep scarlet. Wool is that which appears to have the strongest affinity to colouring matter; next to it is silk; then linen; and cotton the weakest, and is therefore the most difficult of all to dye perfectly. Thus, if a piece of linen cloth be dipped into a solution of madder, it will come out just tinged with the colour; but if a piece of the same be previously dipped into a solution of alum or copperas, and dried previously to being dipped in the madder, the alum will become so far impregnated with the colouring principle, that the cloth will receive a perfect dye, and be so fixed that it cannot be separated by any common means. Thus it will be observed, that the art of dyeing permanent colours depends on this intermediate principle, which is termed a mordant. These mordants are very numerous; and on a knowledge of them appears to rest the principal secret of dyeing. The following mode is, however, a very convenient one for makig experiments on fixing the colouring principles of any vegetable extract: To have several pieces of cloth, woollen, cotton, silk, and linen, dipped in the different mordants, and by keeping a small vessel filled with the colouring solution on a fire in a state a little below boiling, by cutting small pieces of each, and immersing them in the colour, and examining and comparing with each other. Experiments of this kind are well worth the attention of persons; for, when we refer to this department, we shall find very few plants which are either now, or ever have been, cultivated for this purpose, although it is well known that so many contain this principle. I have inserted the following, as being known to contain the different colours mentioned; but there are many other plants equally productive of this principle that remain quite unnoticed at present.

539. ACANTHUS mollis. BEAR'S-BREECH.—This gives a fine yellow, which was in use among the ancients.

540. ACTAEA spicata. BANEBERRY.—The juice of the berries affords a deep black, and is fixed with alum.

541. ANCHUSA officinalis. YELLOW ANCHUSA, or BLUE-FLOWERED BUGLOSS.—The juice of the corolla gives out to acids a beautiful green.

542. ANTHEMIS tinctoria.—The flowers afford a shining yellow.

543. ANTHYLLIS vulneraria. KIDNEY-VETCH.—The whole plant gives out a yellow, which is in use for colouring the garments of the country- people.—Linn.

544. ARBUTUS uva-ursi. BEAR'S-BERRY.—The leaves boiled in an acid will dye a brown.

545. ASPERULA tinctoria. WOODROOF.—The roots give a red similar to madder.

546. ANEMONE Pulsatilla. PASQUE-FLOWER.—The corolla, a green tincture.

547. ARUNDO Phragmites. COMMON REED-GRASS.—The pamicle, a green.

548. BERBERIS vulgaris. BARBERRIES.—The inner bark, a yellow.

549. BROMUS secalinus. BROME-GRASS.—The panicle, a green.

550. BIDENS tripartita. HEMP AGRIMONY..—The herb, a good yellow.

551. BETULA alba. BIRCH.—The leaves, a yellow.

552. BETULA nana. DWARF-BIRCH.—The leaves, a yellow.

553. BETULA Alnus. ALDER.—The bark affords a brown colour; which with the addition of copperas becomes black.

554. CALENDULA officinalis. COMMON MARIGOLD.—The radius of the corolla, if bruised, affords a fine orange. The corolla dried and reduced to powder will also afford a yellow pigment.

555. CALTHA palustris. MARSH-MARIGOLD.—The juice of the corolla, with alum, gives a yellow.

556. CAMPANULA rotundifolia. ROUND-LEAVED BELL-FLOWER.—A blue pigment is made from the corolla; with the addition of alum it produces a green colour.

557. CARPINUS Betulus. HORNBEAM.—The bark, a yellow.

558. CHAEROPHYLLUM sylvestre. COW-PARSLEY.—The umbels produce a yellow colour, and the juice of the other parts of the plant a beautiful green.

559. CARTHAMUS tinctorius. SAFFLOWER.—The radius of the corolla, prepared with an acid, affords a fine rose-coloured tint.

560. CENTAUREA Cyanus. BLUE-BOTTLE.—The juice of the corolla gives out a fine blue colour.

561. COMARUM palustre. MARSH-CINQUEFOIL.—The dried root forms a red pigment. It is also used to dye woollens of a red colour.

562. CUSCUTA europaea. DODDER.—The herb gives out a lightish red.

563. CRATAEGUS Oxycantha. HAWTHORN.—The bark of this plant, with copperas, is used by the Highlanders to dye black.

564. DATISCA cannabina. BASTARD-HEMP.—This produces a yellow; but is not easily fixed, therefore it presently fades to a light tinge.

565. DELPHINIUM Consolida. BRANCHING LARKSPUR.—The petals bruised yield a fine blue pigment, and with alum make a permanent blue ink.

566. FRAXINUS excelsior. MANNA.—The bark immersed in water gives a blue colour.

567. GALIUM boreale. CROSS-LEAVED BEDSTRAW.—The roots yield a beautiful red, if treated as madder.

568. GALIUM verum. YELLOW BEDSTRAW.—The flowers treated with alum produce a fine yellow on woollen. The roots, a good red.

569. GENISTA tinctoria.—The flowers are in use among the country-people for dyeing cloth yellow.

570. GERANIUM sylvaticum. MOUNTAIN CRANESBILL.—The Icelanders use the flowers of this plant to dye a violet colour.

571. HIERACIUM umbellatum. HAWKWEED.—The whole herb bruised and boiled in water gives out a yellow dye.

572. HUMULUS Lupulus. HOP.—The strobiles are used for dyeing; but although they yield a yellow colour, the principal use is as a mordant.

573. HYPERICUM perforatum. PERFORATED ST. JOHN'S WORT.—The flowers dye a fine yellow.

574. IRIS germanica. GERMAN IRIS.—The juice of the corolla treated with alum makes a good permanent green ink.

575. ISATIS tinctoria. WOAD.—The leaves steeped in water till the parts are decomposed, produces a fine blue fecula, which is made into cakes, and sold to the woollen-dyers. For its culture, see p. 32.

576. LICHEN Roccella. ORCHIL.—The fine purple called orchil is extracted from this moss.

577. LITHOSPERMUM officinale. GROMWELL.—The roots afford a fine red, which is used by the young girls in Sweden to colour their faces.

578. LYCOPODIUM complanatum. CLUB-MOSS.—The juice of this plant extracted by an acid forms a most beautiful yellow.

579. LYCOPUS europaeus. WATER-HOREHOUND.—The juice of this gives out a black colour, and is sometimes used by the common people for dyeing woollen cloth. The gypsies are said to use the juice of this plant to colour their faces with.

580. LYSIMACHIA vulgaris. LOOSESTRIFE.—The juice of the whole herb is used to dye woollen yellow.

581. MYRICA Gale. SWEET GALE.—The whole shrub tinges woollen of a yellow colour.

582. NYMPHAEA alba. WHITE WATER-LILY.—The Highlanders make a dye with it of a dark chesnut colour.—Light. Fl. Sc.

583. ORIGANUM vulgare. WILD MARJORAM.—The tops and flowers contain a purple colour, but it is not to be fixed.

584. PHYTOLACCA decandra. VIRGINIAN POKEWEED.—The leaves and berries produce a beautiful rose-colour, but it is very fugacious.

585. PRUNUS domestica. PLUM.—The bark is used by the country people to dye cloth yellow.

586. PYRUS Malus. APPLE,-The bark of this plant, also, produces a yellow colour.

587. QUERCUS Robur. OAK.—The juice of the oak mixed with vitriol forms a black ink; the galls ar employed for the same purpose.

588. RESEDA Luteola. DYER'S WEED, or WELD.—The most usual plant from which the yellow dye is extracted. For its culture, see p. 32.

589. RHAMNUS Frangula. BUCKTHORN.—The bark produces a slight yellow, and the unripe berries impart to wool a green colour.

590. RHAMNUS catharticus. PURGING BUCKTHORN.—The bark yields a most beautiful yellow colour; and the ripe berries in the autumn produce a brilliant scarlet.

591. RHUS Cotinus. VENUS'S SUMACH.—The bark of the stalks produces a yellow colour; the bark of the roots produces a red.

592. RHUS coriaria. ELM-LEAVED SUMACH.—This plant is possessed of the same qualities as the one above.

593. RUBIA tinctorum.—The root produces a red colour. For its culture, see p. 32.

594. RUMEX maritima. DOCK.—The whole herb gives out a yellow colour.

595. SALIX pentandra. WILLOW.—The leaves produce a yellow colour.

596. SCABIOSA succisa. DEVIL'S BIT SCABIUS.—The dried leaves produce a yellow colour.

597. SERRATULA tinctoria. SAW-WORT.—The whole herb produces a yellow tincture.

598. SENECIO Jacobaea. RAGWORT.—The roots, stalks, and leaves, before the flowering season, give out a green colour which can be fixed on wool.

599. STACHYS sylvatica. HEDGE-HOREHOUND.—The whole herb is said to dye a yellow colour.

600. THALICTRUM flavum. YELLOW MEADOW-RUE.—The roots and leaves both give out a fine yellow colour.

601. THAPSIA villosa. DEADLY CARROT.—The umbels are employed by the spanish peasants to dye yellow.

602. TORMENTILLA erecta. ERECT TORMENTIL.—This root is red, and might probably be usefully employed.

603. TRIFOLIUM pratense. MEADOW-CLOVER.—The inhabitants of Scania employ the heads to dye their woollen cloth green.

604. URTICA dioica. NETTLE.—The roots of bettles are used to dye eggs of a yellow colour against the feast of Easter by the religious of the Greek church, as are also madder and logwood for the same purpose.

605. XANTHIUM strumarium. LESSER BURBOCK.—
The whole herb with the fruit dyes a most beautiful yellow.

SECTION XII.—-PLANTS USED IN RURAL OECONOMY.

The following few plants are such as are used for domestic purposes which do not fall under any of the foregoing heads, and I therefore have placed them together here.

606. CONFERVA.—This green thready substance has the power of rendering foetid water sweet; for which purpose, when water is scarce, it is usually put into water-tubs and reservoirs.

607. CORYLUS Avellana. HAZEL NUT.—The young shoots of hazel put into casks with scalding water, render them sweet if they are musty, or contain any bad flavour.

608. CROCUS vernus. SPRING CROCUS.—Is well kown as a spring flower, producing one of the most cheerful ornaments to the flower-garden early in the spring. It affords a great variety in point of beauty and colour, and is an article of considerable trade among the Dutch gardeners, who cultivate a great number of varieties, which every year are imported into this and other countries.

609. EQUISETUM hyemale. DUTCH RUSH.—Of this article great quantities are brought from Holland for the purpose of polishing mahogany. The rough parts of the plant are discovered to be particles of flint.

610. ERIOPHORUM polystachion. COTTON GRASS.—The down of the seeds has been used, instead of feathers, for beds and cushions; and the foliage in the north of Scotland is considered useful as fodder.

611. GALIUM verum. YELLOW LADIES' BEDSTRAW.—The foliage affords the dairy-maid a fine rennet for making cheese.

SECTION XIII.—POISONOUS PLANTS GROWING IN GREAT BRITAIN.

"On the day that thou eatest thereof thou shalt surely die."

I have found it necessary to be particular in my description of the articles in this section, as I find that, although the knowledge of Botany has in some measure increased, yet, in general, we are not better acquainted with the Poisonous Vegetables than we were thirty years ago. Many and frequent are the accidents which occur in consequence of mistakes being made with those plants; but it in general happens that, from feelings easily appreciated, persons do not like to detail such misfortunes; which not only hides the mischief, but prevents, in a great measure, the antidotes becoming so well known as for the good of society we could wish they were. This I experienced in my researches after several facts which I wished to ascertain regarding this subject. However, whilst we have in common use such plants as Foxglove, Hemlock, and Henbane, and which are now so generally sold in our herb-shops, people who sell them ought to be particularly careful not to let such fall into the hands of ignorant persons, and thereby be administered either in mistake or in improper quantities. Our druggists and apothecaries are careful in not selling to strangers the more common preparations of Mercury, or Arsenic, drugs which in themselves carry fear and dismay in their very names; yet we can get any poisonous vegetables either in the common market, or of herb-dealers, which are more likely to be abused in their application than other poisons which are of not more dangerous tendencies.

The effects of Vegetable Poisons on the human frame vary according to circumstances. The most usual are: that of disturbing the nervous function, producing vertigo, faintness, delirium, madness, stupor, or apoplexy, with a consequent loss of understanding, of speech, and of all the senses; and, frequently, this dreadful scene ends in death in a short period.

It is, however, fortunate that these dangerous plants, which either grow wild, or are cultivated in this country, are few in number; and it is not less so, that the most virulent often carry with them their own antidote, as many of them, from their disagreeable taste, produce nausea and sickness, by which their mischief is frequently removed; and when this is not the case, it points out that the best and most effectual one is the application of emetics: and it may be almost considered a divine dispensation, that a plant, very common in all watery places, should be ready at hand, which has from experience proved one of the most active drugs of this nature, and this is the Ranunculus Flammula, Water-Spearwort. The juice of this plant, in cases of such emergency, may be given in the quantity of a table-spoonful, and repeated every three minutes until it operates, which it usually will do before the third is taken into the stomach.

After the vomiting is over, the effects often remain, by part of the deleterious qualities being absorbed by the stomach; and as it often happens, in such cases, that medical assistance may not be at hand, I shall, under the head of each class, give their proper antidote, which should be in all cases applied as soon as possible, even before medical assistance is procured. And it should not be forgotten that, in dreadful cases where the medicine cannot be forced down through the usual channel, recourse should be had to the use of clysters.

Under each of the following heads I shall describe such cases as have come under my notice; as they may be useful for comparison: and shall put under each of the more dangerous the Plantae affines, describing as accurately as possible the differences.

BITTER NAUSEOUS POISONS.

These are much altered by vegetable acids in general, and especially by oxymuriatic acid; but they still retain much of their poisonous quality, which appears to be rendered more active by alkalies. The tanning decoctions of nut-galls, acacia, and other strong astringents, Venice treacle, wine, spiritous liquors, and spices, are useful.

623. CHELIDONIUM majus. CELANDINE.—The yellow juice of this plant is extremely acrid and narcotic. It is not at all like any plant used for culinary purposes, and therefore there is not any great danger likely to arise from its being confounded with any useful vegetable.

624. CICUTA virosa. COWBANE.—Two boys and six girls, who found some roots of this plant in a water-meadow, ate of them. The two boys were soon seized with pain of the pericardia, loss of speech, abolition of all the senses, and terrible convulsions. The mouth closely shut, so that it could not be opened by any means. Blood was forced from the ears, and the eyes were horribly distorted.

Both the boys died in half an hour from the first accession of the symptoms.

The six girls, who had taken a smaller quantity of the roots than the boys, were likewise seized with epileptic symptoms; but in the interval of the paroxysms, some Venice-treacle dissolved in vinegar was given to them; in consequence of which they vomited, and recovered: but one of them had a very narrow escape for her life. She lay nine hours with her hands and feet outstretched, and cold: all this time she had a cadaverous countenance, and her respiration could scarcely be perceived. When she recovered, she complained a long time of a pain in her stomach, and was unable to eat any food, her tongue being much wounded by her teeth in the convulsive fits.

Plantae affines.

Celery is smaller than this plant.

Parsley is also smaller in all its parts.

Alexanders differs from it, as a plant not of so high growth.

Angelica may be mistaken for this, but has a more agreeable scent.

All the water parsneps may be confounded with it: but these are known by the smallness of the umbels; and they are generally in bloom, so that this circumstance is a good criterion.

Care should at all times be taken, not to make use of any umbelliferous plants growing in water, as many of them are, if not altogether poisonous, very unwholesome.

625. COLCHICUM autumnale. MEADOW-SAFFRON.—Baron Stoerch asserts, that on cutting the fresh root into slices, the acrid particles emitted from it irritated the nostrils, fauces, and breast; and that the ends of the fingers with which it had been held became for a time benumbed; that even a single grain in a crumb of bread taken internally produced a burning heat and pain in the stomach and bowels, urgent strangury, tenesmus, colic pais, cephalalgia, hiccup, &c. From this relation, it will not appear surprising that we find several instances recorded, in which the Colchicumproved a fatal poison both to man, and brute animals. Two boys, after eating this plant, which they found growing in a meadow, died in great agony. Violent

symptoms have been produced by taking the flowers. The seeds, likewise, have been known to produce similar effects.

626. OENANTHE crocata. HEMLOCK. WATER DROPWORT.—Eleven French prisoners had the liberty of walking in and about the town of Pembroke; three of them being in the fields a little before noon, found and dug up a large quantity of this plant with its roots, which they took to be wild celery, to eat with their bread and butter for dinner. After washing it a while in the fields they all three ate, or rather tasted of the roots.

As they were entering the town, without any previous notice of sickness at the stomach or disorder in the head, one of them was seized with convulsions. The other two ran home, and sent a surgeon to him. The surgeon first endeavoured to bleed, and then to vomit him; but those endeavours were fruitless, and the soldier died in a very short time.

Ignorant yet of the cause of their comrade's death, and of their own danger, they gave of these roots to the other eight prisoners, who all ate some of them with their dinner: the quantity could not be ascertained. A few minutes after, the remaining two who gathered the plant were seized in the same manner as the first; of which one died: the other was bled, and a vomit forced down, on account of his jaws being as it were locked together. This operated, and he recovered; but he was for some time affected with a giddiness in his head; and it is remarkable, that he was neither sick nor in the least disordered in his stomach. The others being bled and vomited immediately, were secured from the approach of any bad symptoms. Upon examination of the plant which the French prisoners mistook for wild celery, Mr. Howell discovered it to be this plant, which grows very plentifully in the neighbourhood of Haverfordwest.

Although the above account, which Mr. Wilmer has so minutely described, seems well attested, and corroborated by the above gentleman, yet I was informed by the late Mr. Adams, comptroller of the Customs at Pembroke, that the Oenanthe does not, that he could find, grow in that part of the country; but that what the above unfortunate French officers did actually eat was the wild Celery, which grows plentifully in all the wet places near that town. I take the liberty of mentioning this circumstance; as it will serve to keep in mind the fact, that celery, when found wild, and growing in wet places, shold be used cautiously, it being in such situations of a pernicious tendency. For such whose curiosity may lead them to become acquainted with the Oenanthe crocata, it grows in plenty near the Red House in Battersea fields on the Thames' bank. The water-courses on the marsh at Northfleet have great quantities of the Apium graveolens growing in them.

Plantae affines.

Cultivated celery differs from it when young, first in the shape and size of its roots. The Oenanthe is perennial, and has a large root, which on being cut is observed to be full of juice, which exudes in form of globules. The celery, on the contrary, has roots in general much smaller, particularly when in a wild state.

The leaves of celery have somewhat the same flavour, but are smaller; the nerves on the lobes of the leaves are also very prominent, and somewhat more pointed.

When the two plants are in bloom, a more conspicuous difference is apparent in the involucrum and seeds, the character of which should be consulted.

It may be mistaken for Parsley; but it is both much larger in foliage and higher in growth; it is also different from it in the shape of the roots.

These are the two plants most likely to be confounded with it. But the student should also consult the difference existing between this plant and the following, which, although somewhat alike in appearance, may be confounded.

Angelica.

Chervil.

Alexanders.

Hemlock.

Skirret.

Cow Parsley.

Lovage.

Wild Parsnep.

Fool's Parsley.

Hamburgh Parsley.

627. PRUNUS Lauro-cerasus. THE COMON LAUREL.—The leaves of the laurel have a bitter taste, with a flavour

resembling that of the kernels of the peach or apricot; they communicate an agreeable flavour to aqueous and spirituous fluids, either by infusion or distillation. The distilled water applied to the organs of smelling strongly impresses the mind with the same ideas as arise from the taste of peach blossoms or apricot kernels: it is so extremely deleterious in its nature, and sometimes so sudden in its operation, as to occasion instantaneous death; but it more frequently happens that epileptic symptoms are first produced. This poison was discovered by accident in Ireland in the year 1728: before which, it was no uncommon practice there, to add a certain quantity of laurel water to brandy, or other spirituous liquors, to render them agreeable to the palate. At that time three women drank some laurel-water; and one of them a short time afterwards became violently disordered, lost her speech, and died in about an hour.

A gentleman at Guildford, some few years back, also, by making an experiment as he intended on himself, was poisoned by a small dose: he did not survive the taking it more than two hours.

In consequence of the above poisonous principle existing in the laurel, it has been recommended to persons to be cautious hwo they make use of the leaves of that shrub, which is a usual practice with cooks for giving flavour to custards, blanch-mange, and other made-dishes, lest the narcotic principle should be also conveyed, to the detriment of the health of persons who eat of them.

And the same may be said of the kernels of all stone-fruits; for the flavours given to noyau, ratafia, and other liquors which are highly prized by epicures, are all of them derived from the same principle as laurel-water, and which, on chemical investigation, is found to be prussic acid. This exists in considerable quantities in the bitter almond, and which when separated proves to be the most active poison known, to the human as well as all other animal existence. This principle, and its mode of extraction, should not be made more public than the necessity of scientific research requires. We cannot with propriety accuse either this tree or the laurel as being poisonous, because the ingenuity of mankind has found out a mode of extracting this active acidulous principle, and which is so very small in proportion to the wholesome properties of the fruit, as not to be suspected of any danger but for this discovery. As well might we accuse wheat of being poisonous, because it yields on distillation brandy, which has been known to kill many a strong-bodied fellow who has indulged in this favourite beverage to excess. An eminent chemist informs me, that he has made experiments with the oxalic acid, and found that when this was also concentrated, it has similar effects; insomuch that no animal can contain a grain of it if taken into the throat or stomach: and thus might we also be led to consider the elegant, and in itself harmless, wood-sorrel, as a poisonous plant.

ACRID POISONS.

These should be attacked by strong decoctions of oak-bark, gall-nuts, and Peruvian bark; after which soft mucilaginous matters should be used, as milk, fat broth, or emulsions.

628. ACONITUM Napelhus. BLUE MONKSHOOD.— This is a very poisonous plant; and many instances have been adduced of its dangerous effects.

It has probably obtained the name of Wolfsbane, from a tradition that wolves, in searching for particular roots which they in part subsist upon in winter, frequently make a mistake, and eat of this plant, which proves fatal to them.

A weaver in Spitalfields, having supped upon some cold meat and salad, was suddenly taken ill; and when the surgeon employed upon this occcasion visited him, he found him in the following situation:—"He was in bed, with his head supported by an assistant, his eyes and teeth were fixed, his nostrils compressed, his hands, feet, and forehead cold, no pulse to be perceived, his respiration short, interrupted, and laborious."

Soon after he had eaten of the above, he complained of a sensation of heat affecting the tongue and fauces; his teeth appeared loose; and it was very remarkable, although a looking-glass was produced, and his friends attempted to reason him out of the extravagant idea, yet he imagined that his face was swelled to twice its usual size. By degrees the heat, wich at first only seemed to affect the mouth and adjacent parts, diffused itself over his body and extremities: he had an unsteadiness and lassitue in his joints, particularly of the knees and ancles, with an irritable twitching of the tendons, which seemed to deprive him of the power of walking; and he thought that in all his limbs he perceived an evident interruption to the circulation of the blood. A giddiness was the next symptom, which was not accompanied with nausea. His eyes became watery, and he could not see distinctly; a kind of humming noise in his ears continually disturbed him, until he was reduced to the state of insensibility before described.

Plantae affines.

Although the mischief which is recited above occurred from the root having been purchased at market, I do not know of any vegetable in common use likely to be confounded with this. It might by chance be mistaken for the smaller tubers of Jerusalem artichoke.

In foliage it comes near to the other species of Aconitum, and to the perennial Larkspurs.

However, as this is a plant much grown in pleasure-grounds on account of its beautiful blue flowers, great care should be taken not to use any roots taken from such places that cannot be well ascertained.

629. ACONITUM Lycoctonum. YELLOW WOLFSBANE.—Every part of this plant is accounted poisonous. In fact, I think it is proper that all the species should be considered as such, and never be made use of, either in medicine or otherwise, without great care in their administration.

630. ACTAEA spicata. BANEBERRY.—This plant is also considered as a deadly poison; but we have no authentical accounts of its mischievous effects, although Parkinson has mentioned it in these words:—

"The inhabitants of all the mountaines and places wheresoever it groweth, as some writers say, do generally hold it to be a most dangerous and deadly poison, both to man and beast; and they used to kill the wolves herewith very speedily."

This is not a common plant, growing only in some particular situa-tions, as near Ingleborough in Yorkshire.

631. RHUS Toxicodendron. POISON-ASH.-The juice of the leaves of this plant is so very acrid as often to corrode the skin, if the leaves are gathered when the dew is on them. Great care should certainly be taken in the giving such a medicine internally, as also in its preparation, it being usually administered in a dried state.

Planta affinis.

Rhus radicans differs from this in having a more trailing habit of growth; otherwise it is scarcely different, so little so, as to baffle a distinction being made by description alone.

STUPEFYING POISONS.

The substances that deaden the effects of the poisons of this class are vegetable acids, which should be thrown into the stomach in large quantities. After the operation of emetics, cream of tartar is also considered of great use, as also oxymuriatic acid, infusions of nut-gall, oak bark; warm spices are considered also of use, for they may separate some part of the deleterious matter, as is shown by their effect when mixed with decoction of these plants; acerb and astringent wines are also of great use.

632. AETHUSA Cynapium. FOOL'S PARSLEY.—Fool's Parsley seems generally allowed to be a plant which possesses poisonous qualities. Baron Haller has taken a great deal of pains to collect what has been said concerning it, and quotes many authorities to show that this plant has been productive of the most violent symptoms; such as anxiety, hiccough, and a delirium even for the space of three months, stupor, vomiting, convulsions, and death.

Where much parsley is used, the mistress of the house therefore would do well to examine the herbs previous to their being made use of; but the best precaution will be, always to sow that variety called Curled parsley, which cannot be mistaken for this or any other plant. We might also observe, that the scent is strong and disagreeable in the aethusa: but this property, either in the plant or the poison, is not at all times to be trusted in cases of this nature.

Plantae affines.

Parsley. The lobes of the leaves are larger in this plant, and are not quite so deep a green. The leaves of fool's parsley are also finer cleft, and appear to end more in a short point.

Celery, being much larger, cannot easily be confounded with it.

Chervil. Fool's parsley, when young, differs from this plant but very little, being much the same in size, and the laciniae of the leaves of a similar form. Chervil, however, is much lighter in colour, and the flavour more pleasant, both to the taste and smell.

Hemlock is commonly a larger plant; and, exclusive of the generic distinctions, may be generally known by its spotted stalk.

When fool's parsley is in bloom, it is readily known by the length of the involucrum.

633. ATROPA Belladonna. DEADLY NIGHTSHADE.— Some boys and girls perceiving in a garden at Edinburgh the beautiful berries of the deadly nightshade, and unacquainted with their poisonous quality, ate several. In a short time dangerous symptoms appeared; a swelling of the abdomen took place; they became convulsed. The next morning one of them died, and another in the evening of the same day, although all possible care was taken of them.

Another case is related by Dr. Lambert, who was desired to visit two children at Newburn, in Scotland, who the preceding day had swallowed some of the berries of the deadly nightshade. He found them in a deplorable situation. The eldest (ten years of age) was delirious in bed, and affected with convulsive spasms: the younger was not in a much better condition in his mother's arms. The eyes of both the children were particularly affected. The whole circle of the cornea appeared black, the iris being so much dilated as to leave no vestige of the pupil. The tunica conjunctiva much inflamed. These appearances, accompanied with a remarkable kind of staring, exhibited a very affecting scene. The symptoms came on about two hours after they had eaten the berries: they appeared at first as if they had been intoxicated, afterwards lost the power of speaking, and continued the whole night so unruly, that it was with much difficulty they were kept in bed. Neither of these ever recovered.

634. DATURA Stramonium. THORN-APPLE.—The seeds and leaves of the thorn-apple received into the human stomach produce first a vertigo, and afterwards madness. If the quantity is large, and vomiting is not occasioned, it will undoubtedly prove fatal. Boerhaave informs us, that some boys eating some seeds of the thorn-apple which were thrown out of a garden, were seized with giddiness, horrible imaginations, terrors, and delirium. Those that did not soon vomit, died.

635. HYOSCYAMUS niger. HENBANE.—Henbane is a very dangerous poison. The seeds, leaves, and root, received into the human stomach, are all poisonous.

The root in a superior degree produces sometimes madness; and if taken in large quantity, and the stomach does not reject it by vomiting, a stupor and apopleptic symptoms, terminating in death, are the usual consequences.

A case of the bad effects of the roots of this plant, which occurred in Ireland, is mentioned by Dr. Threlkeld. In the winter season, some men working in a garden threw up some roots which were supposed to be Skirrets, and those were cooked for dinner. About two hours after they were eaten, a person who partook of them was taken with an unusual lassitude, as if being much fatigued, heat and dryness both in the mouth an the throat, a giddiness accompanied with dimness of sight, and a partial stoppage in his urine. Several others who had eaten at the same table, as also servants who had partaken, were subjected to the like influence. Medical assistance being at hand, by the use of emetics they were relieved; but it was many days before the whole of them had recovered from those dreadful symptoms.

Two children having both eaten of the berries of this plant, the one a boy (who recovered) being taken ill, vomitted, and was supposed to have thrown them off his stomach: the other, a little girl, died in convulsions the next morning. As mothers and kindred souls do not like names to be made public in these cases, I cannot help feeling some desire to suppress a publicity of a fact in which a near and dear relative was materially interested. In justice, however, to the public, I must mention that I can vouch for the fact, and trust it may not pass without notice, so far as to let the berries be supposed anything but wholesome.

Plantae affines.

The idea of Skirrets being confounded with this plant, is, I think, erroneous, if it has leaves on, as they are not pinnated, and very different from it. When the Hyoscyamus is in bloom, it has curiously-formed flowers of an uncommonly disgusting hue. The scent of this plant, on bruising it, and its general appearance, render it almost impossible that any one should mistake it. The roots, in the winter season, when destitute of leaves, may, however, be mistaken for those of Parsnep, Parsley, Skirret, and many others of similar shape, and of which it is out of our power to give a distinguishing character.

636. LACTUCA virosa. STRONG-SCENTED WILD LETTUCE.—The juice of this plant is a very powerful opiate, and care should be taken how it is made use of. I

have not heard of any dangerous effects having been produced by it. The strong and disagreeable scent and bitter nauseous taste will most likely always operate as a preservative to its being used for food; and as a medicine, it is hoped its use will be confined to the judicious hand of a medical botanist.

Plantae affines.

All the kinds of garden lettuce; but it may be distinguished by its spines on the back of the leaves. It may be remarked, that the milky juice of all lettuce has similar properties to the above; but the juice is not milky till such time as the plant produces seed-stalks, and then the taste in general is too nauseous for it to be eaten.

637. SOLANUM Dulcamara. BITTERSWEET.—The berries of this plant have been sometimes eaten by children, and have produced very alarming effects. It is common in hedges, and should be at all times as much extirpated as possible.

638. SOLANUM nigrum. DEADLY NIGHTSHADE.— Webfer has given us an account of some children that were killed in consequence of having eaten the berries of this plant for black currants. And others have spoken of the direful effects of the whole plant so much, that, from the incontestable proofs of its deleterious qualities, persons cannot be too nice in selecting their pot-herbs, particularly those who make a practice of gathering from dunghills and gardens Fat-Hen, &c. as there is some distant similitude betwixt these plants, and their places of growth are the same.—Curtis's Fl. Lond. fasc. 2.

Plantae affines.

All the Chenopodia grow with this plant wild, and are somewhat alike in appearance; but the Solanum may at all times be distinguished by its disagreeable strong scent.

FOETID POISONS.

These come near to the Stupefying Poisons; but they are not treated in the same manner; for ether, wine, or acids combined with spirits, appear the properest things to destroy their deleterious properties: spices are then indicated, except for savine, which requires instead thereof acids.

639. CONIUM maculatum. HEMLOCK.—Two soldiers quartered at Waltham Abbey collected in the fields adjoining to that town a quantity of herbs sufficient for themselves and two others for dinner when boiled with bacon. These herbs were accordingly dressed, and the poor men ate of the broth with bread, and afterwards the herbs with bacon: in a short time they were all seized with vertigo. Soon after they were comatose, two of them became convulsed, and died in about three hours.

Plantae affines.

Parsley differs from this except in size and colour of the leaves.

Celery is also much like this plant, and particularly so if found wild; but which, for reasons given before, should never be collected to be eaten.

Fool's parsley is very like it; and when the hemlock is in a small state, and this plant luxuriant, I have been in some doubt as to pointing out a perfect difference, especially when they are not in fructification. The spots on hemlock form generally a distinguishing mark.

640. DIGITALIS purpurea. FOXGLOVE.—A few months ago, a child was ill of a pulmonary complaint, and the

apothecary had desired the nurse to procure a small quantity of Coltsfoot and make it a little tea; and accordingly the good woman went to a shop in London, where she procured, as she supposed, three pennyworth of that herb, and made a decoction, of which she gave the patient a tea-cupful; a few minutes after which she found symptoms of convulsions make their appearance, and sent for the apothecary: but who, unfortunately, was so totally ignorant of botany as not to know the plant, but supposing it to be Coltsfoot, after the infant died, took his leave, without ay remark further, than that the disorder which occasioned its death had arisen from some accidental and unusual cause. The nurse, however, did not feel perfectly satisfied of this fact, and carried the remainder of the herb to Apothecaries-Hall; and having applied there for information, was referred to Mr. Leffler, a gentleman who had from his botanical researches that season obtained the Sloanean prize; who told her the mistake. He also went and saw the body, and investigated the whole case in a way that has done that young gentleman great credit; and from him I have been favoured with this account. Had the medical attendant but known the difference between the two plants when he was called in first, there was a chance of the child being saved to its distressed parents. And here was certainly a striking instance of medical men neglecting so far the study of botany, as not to know one of the most useful as well as one of the most dangerous plants of the present Pharmacopoeia.

641. HELLEBORUS foetidus. BEARSFOOT.—The country-people are in the habit of chopping up the leaves of this plant and giving it to children for removing worms; but it is a dangerous medicine, and should be made use of with great caution. It is also recommended as a medicine for the same purpose in horses. As much of the chopped leaves as will lie on a crown-piece, given amongst a feed of corn for

three days, and remitted three days, and repeated thus for nine doses, has been known to remove this disease.

"I heard a melancholy story of a mother in this city; viz. that a Country Colleagh gave some of this plant to her two sons, one of six, the other of four years of age, to kill worms; and that before four in the afternoon they were both corpses."-Dr. Threlkeld, in a short account of the plants in the neighbourhood of Dublin.

642. JUNIPERUS Salvina. SAVINE.—The expressed juice of this plant is very poisonous, and often known to produce the most violent effects. It is sometimes used by persons for expelling worms in children, but should be used with great caution; for, if the quantity taken into the stomach is more than it can digest, all the dreadful effects of the poisons of this class are certain to be the immediate consequence.

643. SCROPHULARIA aquatica. WATER-BETONY.— Every part of this plant is said to be violently narcotic; but its very disagreeable strong scent and extremely bitter taste render it not likely to be used in mistake for any culinary vegetable; and although we know what its effects are from report, we do not think it of so dangerous a tendency as some of our poisonous vegetables.

DRASTIC POISONS.

These purge both upwards and downwards with great violence by means of their acrid poisonous resin, which also violently affects the throat and passages. Although alkalies have been recommended in this case, in order to divide this resin, and that a solution of soap is proper, yet the vegetable acids are also very useful, and have a great effect in diminishing the purgative effect. Besides this, it appears still more advantageous to give astringents: Venice treacle, decoctions of bark or cascarilla, pomegranate rind, and balaustines; all which certainly precipitate this drastic principle.

644. ASCLEPIAS syriaca. SYRIAN DOGSBANE.—All the species of Asclepias have a white acrid juice which is considered poisonous. It is observed to be very acrid when applied to any sensible part of the mouth or throat.

645. BRYONIA alba. WILD VINE, or WHITE BRYONY.—The berries of this plant, when hanging on the hedges, have the appearance of white grapes, and have been eaten by children. They are known to produce dreadful effects; but it frequently happens that they produce nausea on the stomach, by which they operate as an emetic of themselves.

646. EUPHORBIA Lathyris. CAPER SPURGE.—A plant common in old gardens, but not indigenous. The seed-vessels are much in shape of caper-buds: hence its name. People have been in the habit of pickling these berries, from which some dangerous symptoms have arisen; it is probable that the vinegar may have been the means of checking its bad effects. It should, however, never be used as food.

647. EUPHORBIA amygdaloides. WOOD SPURGE.—The juice of this plant has been known to produce very dangerous swellings in the mouth and throat of persons who have occasionally put it into their mouths. We do not know that it is very dangerous; and nothing is likely to tempt any persons to use it as food or otherwise.

648. MERCURIALIS perennis. DOG'S MERCURY.—This plant is of a soporific deleterious nature, and is said to be noxious to both man and beast. Many instances are recorded of its fatal effects.

Mr. Ray acquaints us with the case of a man, his wife, and three children, who were poisoned by eating it fried with bacon: and a melancholy instance is related in the Philosophical Transactions, Number CCIII., of its pernicious effects upon a family who ate at supper the herb boiled and fried. It produced at first nausea and vomiting, and comatose symptoms afterwards; two of the children slept twenty-four hours; when they awoke, they vomited again, and recovered. The other girl could not be awakened during four days; at the expiration of which time she opened her eyes and expired.

Plantae affines.

It appears that the different species of Chenopodium have been mistaken for this plant. I do not see myself any very near likeness: but as all the species of Chenopodium have been called English Mercury, it is possible that the name may have been the cause of the mistake.

649. MERCURIALIS annua. ANNUAL DOG'S MERCURY.—Persons who are in the habit of gathering wild herbs to cook, should be careful of this. It grows plentifully in all rich grounds, and is common with Fat Hen

and the other herbs usually collected for such purposes in the spring, and from which it is not readily distinguished: at least, I cannot describe a difference that a person ignorant of botany can distinguish it by.

650. PERIPLOCA graeca.—This is an ornamental creeping plant, and commonly grown in gardens for covering verandas, and other places for shade.

I once witnessed a distressing case. A nurse walking in a garden gathered flower of this plant, and gave it to a child which she had in her arms. The infant having put it to its mouth, it caused a considerable swelling and inflammation, which came on so suddenly, that, had it not been that one of the labourers had met with a similar accident, no one would have known the cause. The child was several days before it was out of danger, as the inflammation had reached the throat.

651. VERATRUM album. WHITE HELLEBORE.—The roots of this plant, and also of the Veratrum nigrum, have been imported mixed with the roots of yellow gentian, and have proved poisonous.—Lewis's Materia Medica.

POISONOUS FUNGI.

The deleterious effects of these generally show themselves soon after they are in the stomach. Vomiting should be immediately excited, and then the vegetable acids should be given; either vinegar, lemon-juice, or that of apples; after which, give ether and antispasmodic remedies, to stop the excessive bilious vomiting. Infusions of gall-nut, oak-bark, and Peruvian bark, are recommended as capable of neutralizing the poisonous principle of mushrooms. It is however the safest way not to eat any of these plants until they have been soaked in vinegar. Spirit of wine, and ether, extract some part of their poison; and tanning matter decomposes the greatest part of it.

Agaricus bulbosus. ——— necator. ——— mamosus. ——— piperitus. ——— campanulatus. ——— muscarius.

These are kown to be poisonous. But the fungi should all be used with great caution; for I believe even the Champignon and Edible mushroom to possess deleterious qualities when grown in certain places.

SECTION XIV.—PLANTS NOXIOUS TO CATTLE.

The foregoing lists of poisonous plants are most of them of less dangerous tendency to cattle than to the human species: for although many of them may be mistaken for wholesome, yet, when they are growing wild, it will be observed, that the discriminating powers of the brute creation in this point are so correct, that very few have been known to be eaten by them.

The following are a few of a different class, which, as not containing any thing particularly disagreeable to the taste of cattle, are frequently eaten by them to their injury.

The agricultural student should make himself perfectly acquainted with those.

652. CICUTA virosa. COWBANE.—Linnaeus observes, that cattle have died in consequence of eating the roots. It is fortunate that this plant is not very plentiful: it is poisonous to all kinds of cattle except goats. The flower of this plant is not unlike that of water-parsneps, which cows at some seasons will eat great quantities of.

653. BEAR'S GARLICK. Allium ursinum.

654. CROW GARLICK. Allium vineale.

These plants very frequently occur in meadow-land, and have property of giving a strong garlick flavour to the milk yielded by cows that feed there; and which is often also communicated to the butter.

655. DARNELL GRASS. Lolium temulentum.—This grass has the faculty of causing poultry or birds to become intoxicated, and so much so that it causes their death.

656. LOUSEWORT. Pedicularis palustris.—This plant, which abounds in wet meadows, is said to produce a lousy disease in cows if they eat of it.

657. MAYWEED. Anthemis cotula.—This is altogether of such an acrid nature, that the hands of persons employed in weeding crops and reaping, are often so blistered and corroded as to prevent their working. It also has been known to blister the mouths and nostrils of cattle when feeding where it grows.

658. COLCHICUM autumnale. MEADOW-SAFFRON.—This is a common plant in pasture-land in Worcestershire, Herefordshire, and other counties. Many are the instances that have occurred of the bad effects of it to cattle. I have this last autumn known several cows that died in consequence of eating this plant.

659. MELILOT. Trifolium officinale.—This plant when eaten by cows communicates a disagreeable taste to milk and butter.

660. ROUND-LEAVED SUN-DEW. Drosera rotundifolia.—Very common on marshy commons, and is said to be poisonous to sheep, and to give them the disease called the rot.

661. SEA BARLEY-GRASS. Hordeum maritimum.—This grass has been known in the Isle of Thanet and other places to produce a disease in the mouths of horses, by the panicles of the grass penetrating the skin.

662. WATER-HEMLOCK. Phellandrium aquaticum.—Linnaeus informs us that the horses in Sweden by eating of this plant are seized with a kind of palsy, which he supposes is brought upon them, not so much by any noxious qualities in the plant itself, as by a certain insect which breeds in the stalks, called by him for that reason Curculio paraplecticus [Syst. Nat. 510]. The Swedes give swine's dung for the cure.

663. YEW. Taxus baccata.—This is poisonous to cattle: farmers and other persons should be careful of this being thrown where sheep or cattle feed in snowy weather. It is particularly dangerous to deer, for they will eat of it with avidity when it comes in their way.

SECTION XV.—PLANTS NOXIOUS IN AGRICULTURE.

Annual Weeds, or such as grow wild in Fields, and that do not produce any Food for Cattle.

Many weeds are troublesome to the farmer amongst his crops; but which, by affording a little fodder at some season or other, in some degree compensate for their intrusion. But as the following are not of this description, they ought at all times to be extirpated: for it should be recollected, that the space occupied by such a plant would, in many instances, afford room for many ears of wheat, &c.

The following are annuals, and chiefly grow among arable crops, as corn, &c. As these every year spring up from seeds, it is a very difficult matter for the farmer to prevent their increase, especially since the practice of fallowing land has become almost obsolete. It is a fact worthy notice, that the seeds of most of the annual weeds will lie in the ground for many years, if they happen to be place deep: so that all land is more or less impregnated with them, and a fresh supply is produced every time the land is ploughed. It is therefore proper that annual weeds of every description should be prevented as much as possible can be from going to seed, for one year's crop will take several seasons to eradicate. The only effectual mode we are acquainted with of getting rid of annual weeds is, either by hoeing them up when young, or by cutting the plants over with any instrument whilst in bloom; for it should be observed, that those never spring from the roots if cut over at that period of their growth, which oftentimes may be easily accomplished.

I once observed a crop of burnet, in which Bromus secalius (Lob Grass) was growing, whose spike stood a considerable height above the crop, and several acres of which a boy or woman might have cut over in a short space of time: but it was not so: the grass seeds and burnet were suffered to ripen together, and no means could be devised to separate the two when threshed. For this reason the burnet seeds never could find a market, and consequently the trouble of saving it, as well as the crop, was lost to the grower. I mention this as an instance of many that frequently occur. How many times do we see with crops of winter tares wild oats seeding in them? or Carduus mutans standing so high above those crops that they might be thus extirpated with great ease?

It may be observed, that it is in culture of this nature where annual seeds multiply. A regular crop of wheat will, by its thickness on the ground, retard their growth by smothering them; but the other gives them every facility, and particularly autumnal-sown crops.

664. Blue-bottle - - - Centaurea Cyanus.
665. White-blite - - - Chenopodium album.
666. Charlock - - - Sinapis arvensis.
667. Chickweed - - - Alsine media.
668. Cockle - - - Agrostemma Githago.
669. Cleavers - - - Galium Aparine.
670. Corn Marigold - - - Chrysanthemum segetum.
671. Corn Crowfoot - - - Ranunculus arvensis.
672. Corn Chamomile - - - Matricaria Chamomilla.
673. Weak-scented do - - ————— inodora.
674. Grass, Lob - - - Bromus secalinus.
675. ——- Bearded Oat - - Acena fatua.
676. ——- Field Foxtail - Alopecurus agrestis.
677. ——- Darnel - - - Lolium temulentum.
678. Groundsel, common - - Senecio vulgaris.

679. Wall Barley - - - Hordeum murinum.
680. Mallow, common - - - Malva sylvestris.
681. Mayweed, stinking - - Anthemis Cotula.
682. Melilot - - - Trifolium officinale.
683. Mustard, white - - - Sinapis alba.
684. ———-, hedge - - - Erysimum Barbarea.
685. Nettle, Stinging, small - Urtica urens.
686. ———, Dead - - - Lamium albium.
687. Nipplewort - - - Lapsana communis.
688. Orach, wild - - - Atriplex hastata.
689. ——-, spreading - - ——— patulata.
690. Pilewort - - - Ranunculus ficaria.
691. Persicaria, spotted-leaved Polygonum Ficaria.
692. ————, pale-flowered ————- pensylvanicum.
693. ————, climbing - ————- Convolvulus.
694. Pheasant-eye - - - Adonis autumnalis.
695. Poppy, common red - - Papaver Rhoeas.
696. Poppy, long rough-headed - Papaver Argemone.
697. Radish, wild - - - Raphanus Raphanistrum.
698. Shepherd's Needle - - Scandix Pecten Veneris.
699. Spearwort - - - Ranunculus Flammula.
00. Spurry, Corn - - - Spergula arvensis.
701. Thistle, Spear - - - Carduus lanceolatus.
702. ———- Star - - - Centaurea Calcitrapa.
703. ———- Marsh - - - Carduus palustris.
704. ———- Dwarf - - - ———- acaulis.
705. Tine Tare, smooth-podded - Ervum tetraspermum.

Creeping-rooted Weeds.

The following are such as are perennial, and are of the most troublesome nature, being xtremely difficult to get rid of in consequence of their creeping roots. It unfortunately appens that, where the land is the most worked, and the roots the more broken thereby, the more the crop of weeds increases on the land. Therefore, the only effectual mode of extirpating plants of this nature, is by picking out the roots after the plough, or by digging them up at every opportunity by some proper instrument.

Where weeds of this nature occur, there is too often thought to be more labour than profit in their extirpation. And although this is an argument of some propriety, where a farmer is tenant at will, or where his strength is not proportionate to the land: yet if land is worth any thing at all, that, whatever it may be, is lost, if it is suffered thus to become barren. And as prevention is in most cases considered preferable to cure, more care ought to be taken than generally is, of all our hedges and waste pieces of land by road sides, &c. Many of these plants are found growing in such places, and their seeds are of that nature that they are calculated to fly to considerable distances,—a contrivance in nature to fertilize the ground in her own way; but which, as agriculturists, it is the business of men to check.

706. Bindweed, small - - Convolvulus arvensis.
707. Bindweed, large - - ——————- sepium.
708. Bistort - - - Polygonum bistorta.
709. Brakes - - - Pteris aquilina.
710. Clown's Woundwort - - Stachys palustris.
711. Cammock - - - Ononis arvensis.
712. Coltsfoot - - - Tussilago Farfara.
713. Crowfoot, creeping - - Ranunculus repens.

714. Goutweed - - - Aegopodium Podagraria.
715. Grass, Garden Couch - Triticum repens.
716. ——-, Couchy-bent - Agrostis stolonifera.
717. ——-, Couch Oat, or Knot Avena elatior.
718. —-, Creeping-soft - Holcus mollis.
719. Horsetail, Corn - - Equisetum arvense.
720. Persicaria, willow-leaved Polygonum amphibium.
721. Rest Harrow - - - Ononis spinosa.
722. Sow-Thistle, Corn - - Sonchus arvensis.
723. Spatling Poppy - - Cucubalus Behen.
724. Stinging-Nettle, large - Urtica dioica.
725. Silverweed - - - Potentilla anserina.
726. Sneezewort - - - Achillea Ptarmica.
727. Thistle, melancholy - Carduus heterophyllus.
728. ———-, cursed - - ———- arvensis.
729. Water Horehound - - Lycopus europaeus.

Perennial Weeds.

This enumeration of noxious plants contains principally those which, although they are very troublesome, are more easy of extirpation than the last: for although the most of them are perennial, yet, as their roots do not spread as those of the above list do, they are to be effectually removed by taking up the plants by their roots. It should, however, be always noticed, that it is to little account to endeavour to clear any land of such incumbrances, if any waste places which are separated only by a hedge are allowed to grow these things with impunity; for the seeds will invariably find their way. The contrivance of nature in their formation is a curious and pleasant subject for the philosophical botanist; at the same time it is one of those curses which was impelled on human labour.

730. Butter-bur - - - Tussilago Petasites.
731. Burdock - - - Arctium Lappa.
732. Bugloss, small - - Lycopis arvensis.
733. Crowfoot, round-rooted - Ranunculus bulbosus.
734. ————, tall - - Ranunculus acris.
735. Dock, curdled - - - Rumex crispus.
736. ——, broad-leaved - - ——- obtusifolius.
737. ——, sharp-pointed - ——- acutus.
738. Fleabane, common - - Inula dysenteria.
739. Garlick, crow - - - Allium vineale.
740. ————-, bear - - - ———— ursinum.
741. Grass, turfy hair - - Aira caespitosa.
742. ——-, meadow soft - - Holcus lanatus.
743. ——-, carnation - - Carex caespitosa.
744. Knapweed, common - - Centaurea nigra.
745. ————, great - - ————- Scabiosa.
746. Mugwort - - - Artemisa vulgaris.
747. Meadow-sweet - - - Spiraea ulmaria.
748 Ox-eye Daisy Chrysanthemum Leucanthe-mum

749. Plantain, great - - Plantago major.
750. Ragwort, common - - Senecio Jacobaea.
751. ———-, marsh - - ———- aquaticus.
752. Rush, common - - - Juncus conglomeratus.
753. ——, blueish - - - ——— glaucus.
754. ——, flat-jointed - - ——— squarrosus.
755. ——, round-jointed - ——— articulatus.
756. ——, bulbous - - - ——— bulbosus.
757. Scabious, common - - Scabiosa avensis.
758. Thistle, milk - - - Carduus marianus.
759. ———-, meadow - - ———- pratensis.

SECTION XVI.-EXOTIC TREES AND SHRUBS.

The fashionable rage for planting ornamental trees and shrubs having so much prevailed of late years, that we meet with them by the road sides, &c. almost as common as we do those of our native soil, I have therefore enumerated them in this section.

Our limits will not admit of giving any particular descriptions of each; but as persons are often at a loss to know what soil each tree is known to thrive in best, we have endeavoured to supply that information; which will be understood by applying to the following

ABBREVIATED CHARACTERS.

c.m. read common garden mould.
b.m. - bog mould.
 l. - loam.
b.l. - bog and loam, the greater part bog.
l.b. - loam and bog, the greater part loam.
s. - sheltered situation.
a. - annual.
bi. - biennial.
p. - perennial.
shr. - tree or shrub.
c. - creeper.
w. - adapted to covering walls.

As the soils recommended may not be generally understood; a little attention to the following rules will enable persons to discover what is fit for their purposes.

Loam—the kind best adapted to the purpose of growing plants, is of a moderately close texture, between clay and sand, differing from the former in want of tenacity when wet; and not becoming hard when dry; nor is it loose and dusty like the latter; but in both states possesses somewhat of a saponaceous quality. It varies in colour from yellow to brown, and is commonly found in old pastures: it may also be remarked, that where any perennial species of Clover (Trifolium) are found wild, it is almost a certain indication of a fertile loam, and such as contains the proper food of plants in abundance.

Bog-mould—is frequently found on waste lands, where Heaths (Ericae) are produced: it is composed of decayed vegetable matter and white sand. The best sort is light when dry, of a black colour, and easily reduced to powder. Care should be taken to distinguish it from Peat, which is hard when dry, destitute in a great measure of the sand, and mostly of a red colour. This contains in great quantities sulphureous particles and mineral oil, which are known to be highly destructive to vegetation.

The mould formed from rotten leaves is a good substitute for bog-mould if mixed with sand, and is often made use of for the same purposes. These earths should be dug from the surface to the depth of a few inches and laid in heaps, that the roots, &c. contained therein may be decomposed: and before they are used should be passed through a coarse screen, particularly if intended for plants in pots.

As loam has been found to contain the greatest portion of the real pabulum of plants, it has long been used for such as are planted in pots; and the component parts of bog-earth being of a light nature, a mixture of the two in proper proportions will form a compost in which most kinds of plants will succeed. Attention should be paid to the consistence of the loam; as the more stiff it is, the greater portion of the other is necessary.

DIANDRIA MONOGYNIA.

1 JASMINUM officinale. w. Common white Jasmine c.m. 2 ——— v. argen. variegat. w. Silver-striped ditto c.m. 3 ——— v. aureo variegat. w. Gold-striped ditto c.m. 4 — ——— fruticans, w. Yellow ditto c.m. 5 ——— humile, w. Dwarf yellow ditto b.l. 6 Phillyrea media, w. Privet-leaved Phillyrea c.m. 7 ——— v. virgata Twiggy ditto c.m. 8 ——— v. pendula Pendulous ditto c.m. 9 ——— —- oleaefolia Olive-leaved ditto c.m. 10 ——— buxifolia Box-leaved ditto c.m. 10 ——— angustifolia Narrow-leaved ditto c.m. 12 ——— v. rosmarinifolia Rosemary-leaved ditto c.m. 13 ——— brachiata Dwarf ditto c.m. 14 ——— v. latifolia Broad-leaved ditto c.m. 15 ——— v. laevis Smooth broad-leaved ditto c.m. 16 ——— v. spinosa Prickly broad-leaved ditto c.m. 17 — ——— v. obliqua Hex-leaved ditto c.m. 18 Chionanthus virginicus Fringe Tree b.m. 19 Syringa vulgaris Blue lilac c.m. 20 ——— v. alba White ditto c.m. 21 ——— persica Persian ditto c.m. 22 ——— v. lacinita Cut-leaved ditto c.m. 23 ——— latifolia Broad-leaved ditto c.m.

TETRANDRIA MONOGYNIA.

24 Cephalanthus occidentalis Button-wood b.l. 25 Houstonia coccinea Scarlet Houstonia b.l.s. 26 Buddlea globosa Globe-flowered Buddlea b.l.s. 27 Cornus florida Great-flowering Dog-wood c.m. 28 ——— mascula Cornelian Cherry c.m. 29 ——— sericea Blue-berried ditto c.m. 30 ——— alba White-berried ditto c.m. 31 ——— stricta Upright ditto c.m. 32 ——— sibirica Siberian ditto c.m. 33 ——— paniculata Panicled ditto c.m. 34 —

alternifolia Alternate-leaved ditto c.m. 35 ——— v. virescens Green-twigged ditto c.m. 36 Ptelea trifoliata Shrubby Bean-trefoil c.m. 37 Elaeagnus angustifolia Narrow-leaved Oleaster c.m. 38 ——— v. latifolia Broad-leaved ditto c.m.

TETRANDRIA DIGYNIA.

39 Hamamelis virginica Witch Hazel c.m.

TETRANDRIA TETRAGYNIA.

40 Ilex opaca Carolina Holly b.l. 41 ——— v. angustifolia Narrow-leaved ditto b.l. 42 ——— primoides Deciduous ditto b.l. 43 ——— Cassine Dahoon ditto l. 44 ——— vomitoria South Sea Tea Tree l.

PENTANDRIA MONOGYNIA.

45 Azalea pontica Yellow Azalea b.s. 46 ——— nudiflora Red ditto b.s. 47 ——— v. coccinea Scarlet ditto b.s. 48 ——— v. carnea Flesh-coloured ditto b.s. 49 ——— v. alba Early white ditto b.s. 50 ——— v. bicolor Red and white ditto b.s. 51 ——— v. papilionacea Variegated ditto b.s. 52 ——— v. partita Downy ditto b.s. 53 ——— v. aurantia Orange ditto b.s. 54 ——— v. viscosa Late white ditto b.s. 55 ——— v. vittata White striped ditto b.s. 56 ——— v. fissa Narrow petalled ditto b.s. 57 ——— v. floribunda Cluster-flowered ditto b.s. 58 ——— v. glauca Glaucus-leaved ditto b.s. 59 ——— v. scabra Rough-leaved ditto b.s. 60 Lonicera dioica. c. Glaucous Honeysuckle c.m. 61 ——— sempervirens. c. Trumpet ditto l. 62 ——— grata. c. Evergeen Honeysuckle c.m. 63 ——— implexa. c. Minorca ditto l. 64 ——— nigra Black-berried ditto c.m. 65 ——— tatarica Tartarian ditto c.m. 66 ——— pyrenaica Pyrenean ditto c.m. 67 ——— Alpigena Red-

berried ditto c.m. 68 Lonicera caerulea Blue-berried ditto c.m. 69 ———— Symphoricarpos St. Peter's Wort c.m. 70 ———— Diervilla Yellow-flowered Honeysuckle c.m. 71 ———— Caprifolium c. Italian white ditto c.m. 72 ———— — v. rubra c. Italian early red ditto c.m. 73 ———— Periclym. v. serotina c. Late red ditto c.m. 74 ———— v. quercifolia Oak-leaved ditto c.m. 75 ———— v. belgica Dutch ditto c.m. 76 Lycium barbarum. w. Willow-leaved Boxthorn c.m. 77 ———— europaeum. w. European ditto c.m. 78 Sideroxylon lycoides Willow-leaved Iron-wood b.l. 79 Rhamnus latifolius Broad-leaved ditto c.m. 80 ———- alpinus Alpine ditto b.m. 81 ———- theezans Tea ditto c.m. 82 ———- alnifolius Alder-leaved ditto c.m. 83 ———- Paliurus Christ's Thorn c.m. 84 ———- volubilis. c. Supple-jack Tree c.m. 85 ———- Ziziphus Shining-leaved ditto c.m. 86 ———- Alaternus Common Alaternus c.m. 87 ———- fol. argen. var. Silver-striped ditto c.m.s. 88 — ———- fol. aureo var. Gold-striped ditto c.m.s. 89 ———- v. angustifolius Jagged-leaved ditto c.m. 90 Celastrus scandeus Climbing Staff-Tree c.m. 90 Ceanothus americanus New Jersey Tea Tree c.m. 92 Euonymus latifolius Broad-leaved Spindle-Tree c.m. 93 ———— verrucosus Warted ditto c.m. 94 ———— atro-purpureus Purple-flowered ditto c.m. 95 ———— americanus Evergreen ditto c.m. 96 Itea virginica Virginian Itea b.l. 97 ———- buxifolia Box-leaved ditto b.l. 98 Ribes glandulosum Glandulous Currant c.m. 99 ———- petraeum Rock ditto c.m. 100 ———- floridum Large-flowered ditto c.m. 101 ———- diacanthum Two-spined Gooseberry c.m. 102 ———- oxyacanthoides Hawthorn-leaved ditto c.m. 103 ———- canadense Canadian ditto c.m. 104 ———- Cynosbatea Prickly-fruited Currant c.m. 105 ———- prostratum Procumbent ditto c.m. 106 ———- alpinum Alpine ditto c.m. 107 Hedera quinquefolia. w. Virginian Creeper c.m. 108 — ———- Helix v. latifolia Broad-leaved Ivy. c. c.m. 109 Vitis vitifera. c. Common Grape c.m. 110 ———- Labrusca. c.

Downy-leaved ditto c.m. 111 ——- vulpina. c. Fox Grape c.m. 112 ——- laciniata. c. Parsley-leaved Vine c.m. 113 ——- arborea. c. Pepper Vine c.m.

PENTANDRIA DIGYNIA.

114 Periploca graeca. c. Virginian Silk-Tree c.m. 115 Salsola prostrata Trailing Saltwort c.m. 116 Ulmus americana American Elm c.m. 117 ——- v. alba White American ditto c.m. 118 ——- v. pendula Drooping ditto c.m. 119 ——- nemoralis Twiggy ditto c.m. 120 ——- pumila Dwarf ditto c.m. 121 ——- crispa Curled-leaved ditto c.m. 122 Bupleurum fruticosum Shrubby Hare's-ear c.m.

PENTANDRIA TRIGYNIA.

123 Rhus Typhinum Virginian Sumach c.m. 124 —— glabrum Smooth ditto c.m. 125 —— Vernix Varnish Tree c.m. 126 —— copallinum Lentiscus-leaved Sumach c.m. 127 —— radicans. c. Upright Poison Ash c.m. 128 —— Toxicodendron. c. Trailing or officinal ditto c.m. 129 —— Cotinus Venus's Sumach c.m. 130 —— Coriaria Elm-leaved ditto c.m. 131 Viburnum Tinus Laurustinus c.m. 132 ———— fol. variegat. Striped-leaved ditto c.m. 133 —— —— lucidum Shining-leaved ditto c.m. 134 ———— strictum Upright ditto c.m. 135 ———— nudum Oval-leaved Viburnum c.m. 136 ———— cassinoides Thick-leaved ditto l.s. 137 ———— nitidum Shining-leaved ditto b.l. 138 ———— laevigatum Cassioberry Bush b.l. 139 —— ———— prunifolium Thick-leaved Viburnum c.m. 140 —— —— Lentago Pear-leaved ditto c.m. 141 ———— dentatum Tooth-leaved ditto c.m. 142 ———— v. pubescens Downy-leaved ditto c.m. 143 ———- - acerifolium Maple-leaved ditto c.m. 144 ———— Opulus v. americana American Gelder Rose c.m. 145 ———— v.

rosea Snow-ball ditto c.m. 146 ——— alnifolium Alder-leaved ditto c.m. 147 Sambucus canadensis Canadian Elder c.m. 148 ——— nigra v. laciniata Cut-leaved ditto c.m. 149 ——— racemosa Clustered-flowered ditto c.m. 150 Staphylea trifolia Three-leaved Bladder-Nut c.m. 151 Tamarix germanica German Tamarisk c.m.

PENTANDRIA PENTAGYNIA.

152 Aralia spinosa Angelica Tree b.l.

PENTANDRIA POLYGYNIA.

153 Zanthorhiza Apifolium Parsley-leaved Zanthorhiza b.

HEXANDRIA MONOGYNIA.

154 Prinos verticillatus Whorl-leaved Winter-berry b.l. 155 ——— glaber Smooth ditto b.l. 156 ——— lanceolatus Lanceolate-leaved ditto b.l. 157 ——— laevigatus Spear-leaved ditto b.l. 158 Berberis canadensis Canadian Barberry b.l. 159 ——— cretica Cretan ditto b.l. 160 ——— sibirica Siberian ditto b.l.

HEPTANDRIA MONOGYNIA.

161 Aesculus Hippocastanum Common Horse Chesnut c.m. 162 ——- flava Yellow-flowered ditto c.m. 163 ——- Pavia Scarlet-flowered ditto c.m. 164 ——- parviflora Small-flowered ditto c.m.

OCTANDRIA MONOGYNIA.

165 Koelreuteria paniculata Panicled Koelreuteria b.l. 166 Vaccinium stamineum Green-twigged Bleaberry b.m. 167 ——— diffusum Shining-leaved ditto b.m. 168 ———

—- fuscatum Brown ditto b.m. 169 ————- angustifolium Narrow-leaved ditto b.m. 170 ————- frondosum Obtuse-leaved ditto b.m. 171 ————- venustum Red-twigged ditto b.m. 172 ————- resinosum Clammy ditto b.m. 173 ————- amoenum Broad-leaved ditto b.m. 174 ————- virgatum Twiggy-leaved ditto b.m. 175 ————- tenellum Gale-leaved ditto b.m. 176 ————- macrocarpon Large-fruited ditto b.m. 177 ————- nitidum Shining-leaved ditto b.m. 178 ——— ——- ligustrinum Privet-leaved ditto b.m. 179 ————- pumilum Dwarf ditto b.m. 180 Erica ciliaris Ciliated Heath b.m.s. 181 ——- mediterranea Mediterranean ditto b.m.s. 182 ——- australis Spanish ditto b.m.s. 183 ——- herbacea Herbaceous ditto b.m. 184 ——- arborea Tree ditto b.m.s. 185 Daphne alpina Alpine Daphne b.l. 186 ——— pontica Two-flowered ditto b.l.s. 187 ——— Cneorum Trailing ditto b.l. 188 ——— Tartonraira Silver-leaved Daphne b.l.s. 189 ——— collina Hairy ditto b.l.s. 190 ——— Gnidium Flax-leaved ditto b.l.s. 191 Dirca palustris Marsh Leatherwood b.m.

OCTANDRIA DIGYNIA.

192 Polygonum frutescens Shrubby Polygonum b.s.

ENNEANDRIA MONOGYNIA.

193 Laurus Benzoin Benjamin Tree c.m. 194 ——— nobilis Sweet Bay c.m. 195 Sassafras Sassafras Tree b.l.

DECANDRIA MONOGYNIA.

196 Sophora japonica Japan Sophora c.m. 197 Cercis Siliquastrum European Judas Tree c.m. 198 ——— canadensis American ditto c.m. 199 Guilandina dioica Canadian Bonduc c.m. 200 Ruta graveolens Common Rue

c.m. 201 —— montana Mountain ditto c.m. 202 Kalmia latifolia Broad-leaved Kalmia b.s. 203 —— angustifolia Narrow-leaved ditto b.s. 204 —— v. carnea Pale-flowered ditto b.s. 205 —— glauca Glaucus-leaved ditto b.s. 206 Ledum palustre Marsh Rosemary b.s. 207 —— v. decumbens Dwarf ditto b.s. 208 —— latifolium Labrador Tea b.s. 209 —— buxifolium Box-leaved Ledum b.s. 210 Rhodora canadensis Canadian Rhodora b.m. 211 Rhodorendron ferrugineum Rusty-leaved Rhododendron b.m. 212 —————— dauricum Dauric ditto b.m. 213 — ———— hirsutum Hairy ditto b.m. 214 —————— ponticum Pontic ditto b.m. 215 —————— fol. variegat. Striped-leaved ditto b.m. 216 ——————cataubiense Large ditto b.m. 217 —————— maximum Large-leaved ditto b.m. 218 —————— punctatum Dotted ditto b.m. 219 Andromeda mariana Maryland Andromeda b.m. 220 ————- v. oblonga Oval-leaved ditto b.m. 221 ————- ferruginea Rusty-leaved ditto b.m. 222 ————- polyfolia, v. major Broad-leaved rusty ditto b.m. 223 ————- paniculata Panicled ditto b.m. 224 ————- arborea Tree ditto b.m. 225 ————- racemosa Branching ditto b.m. 226 ——— axillaris Notch-leaved ditto b.m. 227 ————— coriacea Thick-leaved ditto b.m. 228 ————- acuminata Acute-leaved ditto b.m. 229 ————- calyculata Globe-flowered ditto b.m. 230 ————- v. latifolia Broad Box-leaved ditto b.m. 231 ————- v. angustifolia Narrow-leaved ditto b.m. 232 — ————- Catesbaei Catesby's ditto b.m. 233 Epigaea repens Creeping Epigaea b.s. 234 Gualtheria procumbens Procumbent Gualtheria b.s. 235 Arbutus Unedo Common Strawberry Tree b.l. 236 ————- v. fl. rubro Scarlet-flowered ditto b.l. 237 ——–- v. flore pleno Double-flowered ditto b.l. 238 ————- v. angustifolia Narrow-leaved ditto b.l. 239 ————- v. crispa Curled-leaved ditto b.l. 240 ————- Andrachne Eastern ditto b.l. 241 Clethra alnifolia Alder-leaved Clethra b.l. 242 ————- v.

pubescens Pubescent ditto b.l. 243 Styrax officinale Official Styrax b.l. 244 ——— grandifolium Large-leaved ditto l. 245 ——— laevigatum Smooth-leaved ditto l.

DECANDRIA DIGYNIA.

246 Hydrangea arborescens Tree Hydrangea c.m. 247 ——— ——- hortensis Changeable-flowered ditto c.m. 248 ——— ——- glauca Glaucous-leaved ditto b.l. 249 ——————- radiata Rayed-flowered ditto b.l.

DODECANDRIA MONOGYNIA.

250 Halesia tetraptera Wing-seeded Snow-drop Tree c.m.

DODECANDRIA TRIGYNIA.

251 Euphorbia spinosa Shrubby Euphorbia b.l. 252 Aristotelia Macqui Shining-leaved Aristotelia b.s.

ICOSANDRIA MONOGYNIA.

253 Philadelphus coronarius Common Syringa c.m. 254 — —————— nanus Dwarf ditto c.m. 255 Punica Granatum. w. Pomegranata l.w.s. 256 ——— flore pleno. w. Double-flowered ditto l.w.s. 257 ——— flore luteo. w. Yellow-flowered ditto l.w.s. 258 ——— flore albo. w. White-flowered ditto l.w.s. 259 ——— nana. w. Dwarf ditto l.w.s. 260 Amygdalus Persica Peach Tree c.m. 261 —————- v. flore pleno Double-flowering ditto c.m. 262 —————- v. Nectarina Nectarine c.m. 263 —————- nana Rough-leaved Almond c.m. 264 —————- pumila Dwarf ditto c.m. 265 ————- communis Common ditto c.m. 266 — ————- fol. variegat. Striped-leaved ditto c.m. 267 ——— —- chinensis Chinese ditto c.m. 268 ————— orientalis Silvery-leaved ditto c.m. 269 ————— sibirica Siberian

ditto c.m. 270 Prunus virginiana Virginian Bird-Cherry c.m. 271 ——— caroliniana Carolinian ditto c.m. 272 ——— lusitanica Portugal Laurel c.m. 273 Lauro-Cerasus Common Laurel c.m. 274 ———- Maheleb Perfumed Cherry c.m. 275 ———- Armeniaca Apricot Tree c.m. 276 ———- pumila Dwarf Bird-Cherry c.m. 277 ———- pendula Weeping Cherry c.m. 278 ———- pennsylvanica Pennsylvanian Bird-Cherry c.m. 279 ———- nigra Black ditto c.m. 280 ———- cerasifera Mirobalum Plum-Tree c.m. 281 ———- rubra Cornish Bird-Cherry c.m. 282 ———- Cerasus, v. flore pleno Double-flowering ditto c.m. 283 —- domestica Common Plum c.m. 284 ———- v. flore pleno Double-flowering ditto c.m. 285 ———- sibirica Siberian ditto c.m.

ICOSANDRIA DIGYNIA.

286 Crataegus Crus galli Cockspur Thorn c.m. 287 ——— v. pyracanthifolia Pyracanthus-leaved ditto c.m. 288 ——— salicifolia Willow-leaved ditto c.m. 289 ——— Aria, v. suecica Swedish White Beam Tree c.m. 290 ——— coccinea American Hawthorn c.m. 291 ——— sanguinea Bloody ditto c.m. 292 ——— cordata Maple-leaved ditto c.m. 293 ——— pyrifolia Pear-leaved ditto c.m. 294 ——— elliptica Oval-leaved ditto c.m. 295 ——— glandulosa Hollow-leaved ditto c.m. 296 ——— flava Yellow-berried ditto c.m. 297 ——— parviflora Gooseberry-leaved ditto c.m. 298 ——— punctata Great-fruited ditto c.m. 299 ——— v. aurea Great Yellow-fruited ditto c.m. 300 ——— Azarolus Parsley-leaved ditto c.m. 301 ——— monogynia, v. coc. Scarlet Thorn c.m. 302 ——— tomentosa Woolly-leaved ditto c.m. 303 ——— odoratissima Sweet-scented ditto c.m.

ICOSANDRIA PENTAGYNIA.

304 Mespillus Pyracantha Evergreen Thorn c.m. 305 ────── ────- Chamae Mespillus Bastard Quince c.m. 306 ────── ──- canadensis Snowy Service c.m. 307 ─────── Cotoneaster Dwarf Mespilus c.m. 308 ─────── arbutifolia Arbutus-leaved ditto c.m. 309 ──────- fructu rubro Red-fruited ditto c.m. 310 ──────- fructu albo White-fruited ditto c.m. 311 ──────- tomentosa Woolly ditto c.m. 312 ──────- Amelanchier Alpine ditto c.m. 313 ──────- pennsylvanica Pennsylvanian ditto c.m. 314 Pyrus Pollveria Woolly-leaved Pear-tree c.m. 315 ────- spectabilis Chinese Apple-tree c.m. 316 ────- prunifolia Large Siberian Crab c.m. 317 Pyrus baccata Small Siberian Crab c.m. 318 ────- coronaria Sweet-scented ditto c.m. 319 ────- angustifolia Narrow-leaved ditto c.m. 320 ────- Cydonia Common Quince c.m. 321 ────- salicifolia Willow-leaved Crab c.m. 322 ────- praecox Early-flowering ditto c.m. 323 Spiraea laevigata Smooth-leaved Spiraea c.m. 324 ────── salicifolia Willow-leaved ditto c.m. 325 ────── v. paniculata Panicled ditto c.m. 326 ────── ── v. latifolia Broad-leaved ditto c.m. 327 ────── tomentosa Woolly-leaved ditto c.m. 328 ────── Hypericifolia Hypericum-leaved ditto c.m. 329 ────── crenata Crenated ditto c.m. 330 ────── chamaedrifolia Germander-leaved ditto c.m. 331 ────── thalictroides Meadow Rue leaved ditto l. 332 ────── Opulifolia Guelder Rose leaved ditto c.m. 333 ────── sorbifolia Mountain Ash-leaved ditto b.m. 334 ────── sibirica Siberian ditto c.m.

ICOSANDRIA POLYGYNIA.

335 Rosa Lutea Single Yellow Rose l. 336 ────── bicolor Red and Yellow Austrian ditto l. 337 ────── sulphurea Double Yellow ditto l.s. 338 ────── blanda Hudson's Bay ditto l. 339 ────── cinnamonema. fl. pl. Double cinnamon ditto c.m. 340 ────── pimpinellifolia Small Burnet-leaved

ditto c.m. 341 —— spinosissima v. Striped-flowered Scotch Rose c.m. 342 —— v. ruberrima Red Scotch ditto c.m. 343 —— v. flore pleno Double Scotch ditto c.m. 344 —— v. altissima Tall Scotch ditto c.m. 345 —— v. versicolor Marbled Scotch ditto c.m. 346 — carolina Single Burnet-leaved ditto c.m. 347 —— v. flore-pleno Double Burnet-leaved ditto c.m. 348 —— v. pimpinellifolia Single Pennsylvanian ditto c.m. 349 —— v. pimpinellifol. fl. pl. Double Pennsylvanian ditto b.m. 350 —— v. diffusa Spreading Carolina ditto c.m. 351 —— v. stricta Upright Carolina Rose c.m. 352 —— villosa, v. flore pleno Double Apple-bearing ditto c.m. 353 —— provincialis Common Provins ditto c.m. 354 —— v. ruberrima Scarlet Provins ditto c.m. 355 —— v. pallida Blush Provins ditto c.m. 356 —— v. alba White Provins ditto c.m. 357 —— v. multiflora Rose de Meaux c.m. 358 —— v. bicolor Rose de Pompone c.m. 359 —— v. humilis Rose de Rheims c.m. 360 —— v. prolifera Childing's Provins ditto c.m. 361 —— v. lusitanica Blandford or Portugal ditto c.m. 363 —— v. ——————— Rose St. Francis c.m. 363 Rosa provincialis v. —— Shailer's Provins ditto c.m. 364 —— ferox Hedgehog ditto c.m. 365 —— brancteata Ld. Macartney's White Rose c.m. 366 —— centifolia Dutch Hundred-leaved ditto c.m. 367 —— v. rubicans Blush Hundred-leaved ditto c.m. 368 —— v. Singletoniae Singleton's Hundred-leaved do. c.m. 369 —— v. holosericea Single Velvet ditto c.m. 370 —— v. holoserica fl. pl. Double Velvet ditto c.m. 371 —— v. sultana Sultan Rose c.m. 372 —— v. stebennensis Stepney ditto c.m. 373 —— v. ——————— Lisbon ditto c.m. 374 —— v. ——————— Bishop ditto c.m. 375 —— v. ——————— Cardinal ditto c.m. 376 —— v. ——————— Blush Royal ditto c.m. 377 —— v. ——————— Petit Hundred-leaved ditto c.m. 378 —— v. ——————— Pluto ditto c.m. 379 —— v. —— ——————— Monstrous Hundred-leaved do. c.m. 380 —— v. — ——————— Fringe ditto c.m. 381 —— v. ———————

Plicate ditto c.m. 382 —— v. ———————— Two-coloured Hund.-leaved do. c.m. 383 —— v. ———————— Shell ditto c.m. 384 —— parvifolia Burgundy Rose b.m. 385 —— gallica Red officinal Rose c.m. 386 —— v. versicolor Rosa mundi c.m. 387 —— v. marmorea Marbled Rose c.m. 388 —— v. Royal Virgin ditto c.m. 389 —— v. major Giant ditto c.m. 390 —— damascena Red Damask ditto c.m. 391 —— v. rubicans Blush Damask ditto c.m. 392 —— v. versicolor York and Lancaster ditto c.m. 393 —— v. menstrualis Red Monthly ditto c.m. 394 —— v. menstrualis alba White Monthly ditto c.m. 395 —— v. Belgica Blush Belgic ditto c.m. 396 —— v. ———- Great Royal ditto c.m. 397 —— v. ———- Blush Monthly ditto c.m. 398 —— v. ———- Red Belgic ditto c.m. 399 —— v. ———- Goliah Rose c.m. 400 —— v. ———- Imperial Blush ditto c.m. 401 —— multiflora Many-flowered ditto c.m. 402 —— sempervirens. c. Evergreen Rose c.m. 403 —— turbinata Frankfort ditto c.m. 404 —— rubiginosa v. Semidoule Sweet Briar c.m. 405 —— v. muscosa Mossy ditto c.m. 406 —— v. sempervirens Manning's Blush ditto c.m. 407 —— v. flore pleno Double Red ditto c.m. 408 —— v. Royal ditto c.m. 409 —— muscosa Moss Provence Rose c.m. 410 —— moschata Single Musk ditto c.m. 411 Rosa v. flore pleno Double Musk Rose c.m. 412 —— alpina Alpine Rose c.m. 413 —— v. rubro Red Alpine ditto c.m. 414 —— canina, v. flore pleno Double Dog-rose c.m. 415 —— pendulina Rose without Thorns c.m. 416 —— alba Single White Rose c.m. 417 —— v. flore pleno Double White ditto c.m. 418 —— v. prolifera Cluster Maiden's Blush ditto c.m. 419 —— v. major Great Maiden's Blush ditto c.m. 420 —— procera Tall Rose c.m. 421 —— americana American Yellow ditto c.m. 422 Rubus occidentalis American Bramble c.m. 423 ——- odoratus Flowering ditto c.m. 424 ——- fruticosus inermis. c. Bramble without Thorns c.m. 425 ——- v. laciniata. c. Cut-leaved Bramble c.m. 426 ——- v. flore pleno Double-flowered ditto c.m. 427 Calycanthus floridus Carolina Allspice l. 428 ———— v. oblongus Long-leaved ditto l. 429 ———— praecox. w. Early-flowered Chinese ditto l.s.

POLYANDRIA MONOGYNIA.

430 Tilia americana Broad-leaved American Lime c.m. 431 —— v. corallina Red-twigged ditto c.m. 432 —— pubescens Pubescent ditto c.m. 433 —— alba White-leaved ditto c.m. 434 Cistus populifolius Poplar-leaved Cistus l.s. 435 ——— v. minor Small Poplar-leaved ditto l.s. 436 ——— laurifolius Laurel-leaved ditto l.s. 437 —— — Ladaniferus Gum Cistus c.m. 438 ——— monspeliensis Montpellier Cistus l.s. 439 ——— laxus Waved-leaved ditto l.s. 440 ——— salvifolius Sage-leaved ditto l.s. 441 ——— incanus Hoary ditto l.s. 442 ——— albidus White-leaved ditto l.s. 443 ——— crispus Curled-leaved ditto l.s. 444 ——— halimifolius Sea Purslane-leaved ditto l.s. 445 ——— halimifol. v. angustifol. Narrow-leaved Cistus l.s. 446 ——— umbellatus Umbelled-flowered ditto l.s. 447 — —— roseus Red-leaved ditto l.s. 448 ——— marifolius Marum-leaved ditto l.s. 449 ——— Tuberaria Plantain-leaved ditto l.s. 450 ——— apenninus Apennine ditto c.m. 451 ——— mutabilis Changeable ditto l.s.

POLYANDRIA DIGYNIA.

452 Fothergillia alnifolia Alder-leaved Fothergillia b.s.

POLYANDRIA DIGYNIA.

453 Liriodendron Tulipifera Common Tulip Tree c.m. 454 Magnolia grandiflora Laurel-leaved Magnolia b.l.s. 455 — ——— v. obovata Broad-leaved ditto b.l.s. 456 ——— v. lanceolata Long-leaved ditto b.l.s. 457 ——— v.

ferruginea Ferrugineous ditto b.l.s. 458 ———— glauca Swamp ditto b.l.s. 459 ———— acuminata Blue-flowering ditto b.l.s. 460 ———— tripetala Umbrella Tree b.l.s. 461 ———— auriculata Large-leaved ditto b.l.s. 462 ———— purpurea Purple Chinese ditto b.l.s. 463 Annona triloba Trifid-fruited Custard Apple b.l.s. 464 Atragena alpina. c. Alpine Atragena b.l. 465 ———— austriaca. c. Austrian ditto b.l. 466 Clematis cirrhosa. c. Evergreen Virgin's Bower b.l. 467 ———— florida. c. Large-flowered ditto b.l. 468 ———— flore pleno Double ditto c.m. 469 ———— viticella. c. Purple-flowered ditto b.l. 470 ———— v. fl. pleno. c. Double Purple-flowered ditto c.m. 471 ———— crispa. c. Curled-flowered ditto b.l. 472 ———— orientalis. c. Eastern ditto b.l. 473 ———— virginiana. c. Virginian ditto c.m. 474 ———— flammula. c. Sweet-scented ditto c.m.

DIDYNAMIA GYMNOSPERMIA.

475 Teucrium flavum Yellow Teucrium l.s. 476 Satureja montana Winter Savory c.m. 477 Hyssopus officinalis Common Hyssop c.m. 478 Lavandula Spica Lavender c.m. 479 ———— v. flore albo White-flowered ditto c.m. 480 ———— Stoechas French ditto c.m.s. 481 Phlomis fruticosa Jerusalem Sage c.m. 482 Thymus vulgaris Common Thyme c.m. 483 ———— v. fol. variegat. Silver Thyme c.m. 484 ———— vulgaris. latifolia Broad-leaved Thyme c.m. 485 ———— Zygis Linear-leaved ditto c.m.

DIDYNAMIA ANGIOSPERMIA.

486 Bignonia Catalpa Common Catalpa c.m. 487 ———— radicans Great trumpet Flower c.m. 488 ———— v. minor Small ditto c.m. 489 ———— capreolata Four-leaved ditto l.s. 490 Vitex Agnus Castus Chaste Tree c.m. 491 ———— v. latifolia Broad-leaved ditto c.m.

TETRADYNAMIA SILICULOSA.

492 Vella Pseudo-cytisus Shrubby Vella l.s.

MONADELPHIA POLYANDRIA.

493 Hibiscus Syriacus Althaea Frutex c.m. 494 ——— v. ruber Red-flowered ditto c.m. 495 ——— v. albus White-flowered ditto c.m. 496 ——— v. fol. variegat. Striped-leaved ditto c.m. 497 ——— v. flore pleno Double White-flowered ditto c.m. 498 Stuartia Malacodendron Common Stuartia b.l.s. 499 ——— marilandia Maryland ditto b.l.s. 500 Gordonia pubescens Loblolly Bay b.l.s.

DIADELPHIA OCTANDRIA.

501 Polygala Chamaebuxus Box-leaved Milkwort b.m.

DIADELPHIA DECANDRIA.

502 Spartium Junceum Spanish Broom c.m. 503 ——— flore pleno Double-flowered ditto l.s. 504 ——— decumbens Trailing Broom c.m. 505 ——— Scorpius Scorpion ditto c.m. 506 ——— multiflorum Portugal White ditto c.m. 507 ——— patens Woolly-podded ditto c.m. 508 ——— purgans Purging ditto c.m. 509 ——— radiatum Starry ditto b.m. 510 Genista candicans Evergreen genista c.m. 511 ——- triquetra Triangular ditto c.m. 512 ——- sagittalis Jointed ditto l. 513 ——— - sibirica Siberian ditto c.m. 514 ——— - germanica German ditto l. 515 ——— - hispanica Spanish ditto l. 516 ——— - lusitanica Portugal ditto l. 517 Amorpha fruticosa Bastard Indigo c.m. 518 Ononis rotundifolia Round-leaved Rest-Harrow l. 519 ——— fruticosa Shrubby ditto l. 520 Glycine frutescens Shrubby Kidney-bean Tree c.m. 521

Cytisus Laburnum Common Laburnum c.m. 522 ———- v. latifolium Scotch ditto c.m. 523 ———— alpinus Alpine Cytisus c.m. 524 ———— nigricans Black ditto c.m. 525 — ——- divaricatus Divaricated ditto c.m. 526 ———- sessifolius Sessile-leaved ditto c.m. 527 ———— hirsutus Hairy Evergreen ditto c.m.s. 528 ———— purpureus Purple-flowered ditto b.l. 529 ———— austriacus Austrian ditto l. 530 ————- supinus Trailing ditto l. 531 ———- capitatus Large Yellow-flowered ditto c.m. 532 ———- biflorus Two-flowered ditto c.m. 533 Robinia Pseudo-Acacia Common Acacia c.m. 534 ————- hispida Rose Acacia c.m. 535 Robinia glutinosa Glutinous Acacia c.m. 536 ———— Caragana Caragana ditto c.m. 537 ———- Altagana Siberian ditto l. 538 ———— Chamlagu Shining-leaved ditto l. 539 ———— spinosa Thorny ditto l. 540 — ——- Halodendron Salt Tree l. 541 ———— frutescens Shrubby Robinia l. 542 ———— pygmea Dwarf ditto l. 543 ————- jubata Bearded ditto l. 544 Colutea arborescens Common Bladder Senna c.m. 545 ———— cruenta Eastern ditto c.m. 546 ————- Pococki Pocock's ditto c.m. 547 Coronilla Emeris Scorpion Senna c.m. 548 Astralagus tragacantha Goat's Thorn l.

POLYADELPHIA POLYANDRIA.

549 Hypericum calycinum Great-flowered St. John's-wort c.m. 550 ————- hircinum Foetid ditto c.m. 551 ——— —- v. minus Lesser Foetid ditto c.m. 552 ————- elatum Tall ditto c.m. 553 ————- prolificum Proliferous ditto c.m. 554 ———— olympicum Olympian ditto l.s. 555 ————- Kalmianum Kalmia-leaved ditto c.m.

SYNGENESIA POLYGAMIA AEQUALIS.

556 Santolina Chamaecyparissus Lavender cotton c.m. 557 ——————- rosmarinifolius Rosemary-leaved ditto c.m.

SYNGENESIA POLYGAMIA SUPERFLUA.

558 Gnaphalium Stoechas Narrow-leaved Everlasting l.s. 559 Baccharis halimifolia Groundsel tree c.m. 560 Cineraria maritima Sea Rag-wort l.s.

GYNANDRIA PENTANDRIA.

561 Passiflora caerulea. c. Blue Passion Flower c.m.s.

GYNANDRIA HEXANDRIA.

562 Aristolochia Sipho. c. Tree Birthwort l.

MONOECIA TRIANDRIA.

563 Axyris Ceratoides Shrubby Axyris l.s. 564 Comptonia asplenifolia Fern-leaved Gale b.s.

MONOECIA TETRANDRIA.

565 Aucuba japonica Blotched-leaved Aucuba l.b.s. 566 Betula populifolia Poplar-leaved Birch c.m. 567 ——— nigra Black ditto c.m. 568 ——— papyracea Paper ditto c.m. 569 ——— pumila Hairy-leaved Dwarf ditto b.m. 570 ——— oblongata Oblong-leaved ditto c.m. 571 — laciniata Cut-leaved Alder c.m. 572 ——— incana Glaucous-leaved Alder c.m. 573 ——— v. angulata Elm-leaved ditto c.m. 574 Buxus balearicus Minorca Box l.s. 575 ——- semperv. v. variegat. Striped-leaved ditto c.m. 576 ——- v. angustifolia Narrow-leaved ditto c.m. 577 Morus alba White Mulberry c.m. 578 ——- nigra Black

ditto c.m. 579 ——- papyracea Paper ditto c.m. 580 ——- rubra Red ditto c.m.

MONOECIA PENTANDRIA.

581 Iva frutescens Bastard Jesuit's-Bark Tree c.m.

MONOECIA POLYANDRIA.

582 Quercus Phellos Willow-leaved Oak l. 583 ———- v. serioea Dwarf Willow-leaved ditto l. 584 ———- Ilex Evergreen Oak c.m. 585 ———- v. serrata Sawed-leaved Evergreen ditto c.m. 586 ———- v. oblonga Oblong-leaved Evergreen do. c.m. 587 ———- Suber Cork tree c.m. 588 ———- virens Live Oak c.m. 589 ———- Prinos Chesnut-leaved Oak l.s. 590 ———- v. oblonga Long-leaved ditto l. 591 ———- aquatica Water Oak l. 592 ——- v. heterophylla Various-leaved Water Oak l. 593 ——- v. elongata Long-leaved Water ditto l. 594 ———- v. indivisa Entire-leaved Water ditto l. 595 ———- v. attenuata Narrow-leaved Water ditto l. 596 ———- nigra Black Oak c.m. 597 ———- rubra Red ditto c.m. 598 ——- v. coccinea Scarlet ditto c.m. 599 ———- v. montana Mountain Red ditto c.m. 600 ———- discolor Downy-leaved ditto c.m. 601 ———- alba White Oak c.m. 602 ——- aegilops Large prickly-cupped ditto l. 603 ———- Cerris Turkey Oak c.m. 604 Fagus pumila Chinquapin Chesnut l.s. 605 ——- ferruginea Copper Beech c.m. 606 ——- sylvatica v. purpurea Purple ditto c.m. 607 ——- v. asplenifolia Fern-leaved ditto c.m. 608 Carpinus virginiana Virginian Hornbeam c.m. 609 Carpinus Ostrya Hop Hornbeam c.m. 610 Corylus rostrata American Cuckold Nut c.m. 611 ———- Colurna Constantinople ditto c.m. 612 Platanus orientalis Palmated Plane Tree c.m. 613 ——- v. acerifolia Maple-leaved ditto c.m. 614 ———- v. undulata Waved-leaved ditto c.m. 615 ———-

occidentalis Lobed-leaved ditto c.m. 616 Liquidamber Styraciflua Maple-leaved Gum Tree l.

MONOECIA MONADELPHIA.

617 Pinus pinaster Pinaster c.m. 618 ——- Inops Jersey Pine l. 619 ——- resinosa Pitch ditto l. 620 ——- halepensis Aleppo Pine l. 621 ——- Pinea Stone Pine l. 622 ——- Taeda Frankincense ditto l. 623 ——- v. rigida Three-leaved ditto l. 624 ——- v. variabilis Two and three-leaved ditto l. 625 ——- v. alopecuroides Fox-tail ditto l. 626 ——- v. Cembra Siberian stone ditto c.m. 627 ——- Strobus Weymouth ditto c.m. 628 ——- Cedrus Cedar of Lebanon c.m. 629 ——- Larix Red Larch c.m. 630 ——- v. pendula Black Larch c.m. 631 ——- Picea Silver Fir c.m. 632 ——- Balsamea Balm of Gilead Fir c.m. 633 ——- canadensis Hemlock Spruce Fir c.m. 634 ——- nigra Black ditto c.m. 635 ——- alba White ditto c.m. 636 ——- Abies Red or Common ditto c.m. 637 ——- sylvestris v. tatarica Tartarian Pine l. 638 ——- v. montana Mountain ditto l. 639 ——- v. divaricata Hudson's Bay ditto l. 640 ——- v. maritima Sea Pine l. 641 Thuja occidentalis American Arbor-Vitae c.m. 642 ——- orientalis Chinese ditto c.m. 643 Cupressus sempervirens Upright Cypress c.m. 644 —--- v. horizontalis Male Spreading ditto c.m. 645 ——— disticha Deciduous ditto c.m. 646 ——— v. nutans Long-leaved Deciduous ditto l. 647 ——— thyoides Arbor-Vitae-leaved ditto c.m. 648 ——— pendula Cedar of Goa l.s.

DIOECIA DIANDRIA.

649 Salix phylicaefolia Phylica-leaved Willow c.m. 650 —--- babylonica Weeping Willow c.m. 651 ——- retusa Blunt-leaved ditto c.m. 652 Salix incubacea Spreading Willow c.m. 653 ——- ulmifolia Elm-leaved ditto c.m. 654

―― hastata Halbert-leaved ditto c.m. 655 ―― myrtilloides Myrtle-leaved ditto c.m. 656 ―― Lapponum Lapland ditto c.m. 657 ―― tristis Narrow-leaved American ditto c.m.

DIOECIA TRIANDRIA.

658 Empetrum rubrum Red Crow Berry b.m. 659 Hippophaë canadensis Canada Sea Buck-thorn b.l.s. 660 Myrica cerifera Candleberry Myrtle b.l. 661 ―― v. latifolia Broad-leaved ditto b.l.

DIOECIA PENTANDRIA.

662 Pistachia Terebinthus Pistachia Nut Tree l.s. 663 Xanthoxylum Clava Herculis Tooth-ach Tree c.m.

DIOECIA HEXANDRIA.

664 Smilax aspera. c. Rough Bindweed l.b. 665 ―― lanceolata. c. Spear-leaved ditto l.b. 666 ―― rotundifolia. c. Round-leaved ditto l.b. 667 ―― Bona Nox. c. Ciliated ditto l.b. 668 ―― laurifolia. c. Laurel-leaved ditto l.b. 669 ―― sassaparilla. c. Sassaparilla ditto l.b. 670 ―― tamnoides. c. Briony-leaved ditto l.b. 671 ―― caduca. c. Deciduous ditto l.b.

DIOECIA OCTANDRIA.

672 Populus dilatata Lombardy Poplar c.m. 673 ―― balsamifera Tacamahac ditto c.m. 674 ―― candicans White-leaved ditto c.m. 675 ―― laevigata Smooth-leaved ditto c.m. 676 ―― monilifera Canadian ditto c.m. 677 ―― graeca Athenian ditto c.m. 678 ―― heterophylla Various-leaved ditto c.m. 679 ―― angulata Carolina ditto c.m.

DIOECIA DECANDRIA.

680 Coriaria myrtifolia Myrtle-leaved Sumach c.m.

DIOECIA DODECANDRIA.

681 Menispermum canadense. c. Canada Moon-seed l.b.
682 ——————- carolinianum. c. Carolina ditto l.b.

DIOECIA MONADELPHIA.

683 Juniperus thuifera Spanish Juniper c.m. 684 —————- Sabina Common Savin c.m. 685 —————- v. tamariscifolia Tamarisk-leaved ditto c.m. 686 Juniperus v. fol. variegat. Variegated Savin c.m. 687 —————- virginiana Red Cedar c.m. 688 —————- repens Creeping ditto c.m. 689 —————- Oxycedrus Brown-berried ditto l.b.s. 690 —————- phoenicea Phoenicean ditto l.b.s. 691 —————- bermudiana Bermudian ditto l.b.s. 692 ——— —- communis v. suecica Swedish ditto c.m. 693 —————- montana Alpine ditto l.b. 694 Ephedra monostachya Shrubby Horse tail l.b. 695 ———— distachya Greater ditto l.b. 696 Cissampelos smilacina Smilax-leaved Cissampelos l.b.

DIOECIA SYNGENESIA.

697 Ruscus Hypoglossum Broad-leaved Alexandrian Laurel c.m. 698 ——— Hypophyllum Double-leaved ditto b.m. 699 ——— racemosus Common ditto b.m.

POLYGAMIA MONOECIA.

700 Atriplex Halimus Sea Purslane c.m. 701 Acer tataricum Tartarian Maple c.m. 702 ——— rubrum Scarlet ditto c.m. 703 ——— v. pallidum Pale ditto c.m. 704 ———

saccharinum Sugar Maple c.m. 705 —— platanoides Plane-leaved ditto c.m. 706 —— v. laciniatum Cut-leaved ditto c.m. 707 —— montanum Mountain ditto c.m. 708 —— pensylvanicum Pennsylvanian ditto c.m. 709 —— monspessulanum Montpellier ditto c.m. 710 —— creticum Cretan ditto c.m. 711 —— Negundo Ash-leaved ditto c.m. 712 —— Opalus Italian ditto c.m.

POLYGAMIA DIOECIA.

713 Gleditsia triacanthos Three-thorned Acacia c.m. 714 ——————- v. horrida Strong-spined ditto c.m. 715 ————— v. monosperma Single-seeded ditto c.m. 716 Fraxinus rotundifolia Round-leaved Ash c.m. 717 ————— excelsior v. crispa Curled-leaved ditto c.m. 718 ——————- v. diversifolia Various-leaved ditto c.m. 719 ————— v. pendula Weeping Ash c.m. 720 ————— v. striata Striped-barked ditto c.m. 721 ————— v. variegata Blotch-leaved ditto c.m. 722 ————— Ornus Flowering ditto c.m. 723 ————— americana American ditto c.m. 724 ——————- chinensis Chinese ditto c.m. 725 ————— rotundifolia Round-leaved ditto c.m. 726 Diospyrus Lotus Date Plum Tree c.m. 727 Diospyrus virginiana Virginian Plum Tree c.m. 728 Nyssa integrifolia Mountain Tupello l.b. 729 ——- denticulata Water ditto l.b.

POLYGAMIA TRIOECIA.

730 Ficus Garica Common Fig-Tree c.m.

SECTION XVII.- FOREIGN HARDY HERBACEOUS PLANTS.

In enumerating the foregoing, as well as the plants of the present section, I have had more than one object in view; being desirous to put in only such plants as were ornamental or curious, at the same time to insert none but what are perfectly hardy; yet, independently of this, to make it sufficiently general, to give to such persons who might wish to study plants scientifically, a sufficient number for examples in every genus. For this purpose I have retained a portion of the Umbelliferous and other plants. Although not to be distinguished for their general beauty or appearance, yet they are calculated to afford the student the best plants for comparison, and for that reason I have arranged them according to the Linnaean System.

DIANDRIA MONOGYNIA.

1 Veronica sibirica Siberian Speedwell c.m. 2 ———— virginica Virginian ditto c.m. 3 ———— spuria Bastard ditto c.m. 4 ———— maritima Blue-flowered Sea ditto c.m. 5 ———— longifolia Long-leaved ditto c.m. 6 ———— —— incana Hoary ditto c.m. 7 ———— incicisa Cut-leaved ditto c.m. 8 ———— Allioni Creeping ditto c.m. 9 ———— Teucrium Hungarian ditto c.m. 10 ———— urticaefolia Nettle-leaved ditto c.m. 11 ———— orientalis Oriental ditto c.m. 12 ———— candida White-leaved ditto c.m. 13 ———— multifida Multifid ditto c.m. 14 ———– latifolia Broad-leaved ditto c.m. 15 Verinoca prostrata Trailing Sea Speedwell c.m. 16 ———— austriaca Austrian ditto c.m. 17 ———— pinnata Wing'd-leaved ditto c.m. 18 ———— paniculata Panicled ditto c.m. 19 — ———— Gentianoides Gentian-leaved ditto c.m. 20 Gratiola officinalis Hedge-Hyssop c.m. 21 Verbena urticaefolia

Nettle-leaved Vervain c.m. 22 Lycopus virginicus Virginian Lycopus c.m. 23 Monarda fistulosa Hollow-stalked Monarda l. 24 ——— - didyma Scarlet ditto l. 25 — —— - purpurea Purple ditto l. 26 Salvia lyrata Lyre-leaved Sage l.b. 27 ——— virgata Twiggy-branched ditto c.m. 28 ——— sylvestris Spotted-stalked ditto c.m. 29 ——— nemorosa Spear-leaved ditto c.m. 30 ——— austriaca Austrian ditto c.m. 31 ——— Disermas Long-spiked ditto c.m. 32 ——— verticillata Whorl-flowered ditto c.m. 33 — —— glutinosa Yellow-flowered ditto c.m. 34 ——— lineata Flax-leaved ditto l.b. 35 Collinsonia canadensis Nettle-leaved Collinsonia c.m.

TRIANDRIA MONOGYNIA.

36 Valeriana Phu Garden Valerian c.m. 37 Ixia chinensis Chinese Ixia l.b. 38 Galdiolus communis Common red Corn-Flag c.m. 39 ——————- byzantinus Larger ditto c.m. 40 Iris susiana Chalcedonian Iris l.b. 41 ——— florentina Florentine ditto c.m. 42 —— germanica German ditto c.m. 43 —— lurida Dingy ditto c.m. 44 —— sambucina Elder-scented ditto c.m. 45 —— dalmatica Dalmatian ditto c.m. 46 —— variegata Variegated-flowered ditto c.m. 47 —— biflora Two-flowered ditto l.b. 48 —— pumila Dwarf ditto c.m. 49 —— sibirica Siberian ditto c.m. 50 —— squalens Brown-flowered ditto c.m. 51 —— versicolor Various coloured ditto c.m. 52 —— spuria Spurious ditto c.m. 53 —— ochroleuca Pale Yellow ditto c.m. 54 —— graminea Grass-leaved ditto c.m. 55 —— ephium Spanish Bulbous ditto c.m. 56 —— ephioides English Bulbous ditto c.m. 57 —— persica Persian ditto l.b. 58 —— halophila Long-leaved ditto c.m. 59 —— subbiflora One- and Two-flowered ditto c.m. 60 —— virginica Virginian ditto c.m. 61 Iris aphylla Naked-stalked Iris c.m. 62 —— flexuosa Bending-stalked ditto c.m. 63 Commelina erecta Upright Commelina c.m.

TETRANDRIA MONOGYNIA.

64 Scabiosa alpina Alpine Scabious c.m. 65 ——— leucantha Snowy ditto c.m. 66 ——— sylvatica Broad-leaved ditto c.m. 67 ——— ochroleuca Pale white ditto c.m. 68 Crucianella anomala Anomalous Crucianella c.m. 69 Asperula Taurina Broad-leaved Woodroof c.m. 70 Plantago maxima Broad-leaved Plantain c.m. 71 ——— v. rosea Rose ditto c.m. 72 ——— altissima Tall ditto c.m. 73 ——— asiatica Asiatic ditto c.m. 74 Sanguisorba media Short-spiked Burnet-saxifrage c.m. 75 ——— canadensis Canadian ditto c.m.

PENTANDRIA MONOGYNIA.

76 Anchusa angustifolia Narrow-leaved Bugloss c.m. 77 Pulmonaria angustifolia Narrow-leaved Lungwort l.b. 78 ——— virginica Virginian ditto l.b. 79 Borago orientalis Eastern Borage l.b. 80 Symphytum orientale Eastern Comfrey l.b. 81 ——— asperrimum Siberian ditto c.m. 82 Hydrophyllum virginicum Virginian Water-leaf l.b. 83 ——— canadense Canadian ditto l.b. 84 Lysimachia Ephemeron Willow-leaved Loose-strife l. 85 ——— stricta Bulb-bearing ditto b.s. 86 ——— ciliata Ciliated ditto c.m. 87 Plumbago europaea European Lead-wort c.m. 88 Phlox paniculata Panicled Lychnidea c.m. 89 ——- undulata Wave-leaved ditto c.m. 90 ——— suaveolens White-flowered ditto c.m. 91 ——- carolina Carolina ditto c.m. 92 ——- maculata Spotted-stalked ditto c.m. 93 ——- glaberrima Smooth-stalked ditto c.m. 94 Convolvulus americanus American Bind-weed c.m. 95 Polemonium reptans Creeping Greek Valerian c.m. 96 Campanula persicifolia Peach-leaved Campanula l. 97 ——— ——- pyramidalis Pyramidal ditto l. 98 ——— - lilifolia Lily ditto c.m. 99 ———— - rapunculoides Nettle-leaved

ditto c.m. 100 ———— americana American ditto l. 101 ———— versicolor Various-coloured ditto l.b. 102 —— —— sibirica Siberian ditto l.b. 103 Phyteuma spicata Spike-flowered Horn-Rampion c.m. 104 Triosteum perfoliatum Fever Wort l.b. 105 Verbascum ferrugineum Rusty-leaved Mullein l. 106 ———— phoeniceum Purple-flowered ditto l. 107 Hyoscyamus Scopolia Nightshade-leaved Henbane b. 108 Physalis Alkekengi Winter Cherry c.m. 109 Atropa Mandragora Mandrake l.s. 110 Viola montana Mountain Violet c.m. 111 Tabernamonta Amsonia Alternate-leaved Taberna montana 112 ———————— angustifolia Narrow-leaved ditto l.s.

PENTANDRIA DIGYNIA.

113 Apocynum venetum Spear-leaved Dog's-bane c.m. 114 ———— androsaemifolium Fly-catching ditto l.b. 115 — ———— cannabium Hemp-leaved ditto c.m. 116 Asclepius syriaca Syrian Swallow-wort c.m. 117 ————- amoena Oval-leaved ditto c.m. 118 ————- incarnata Flesh-coloured ditto c.m. 119 ————- sibirica Siberian ditto l.b. 120 ————- Vincetoxicum Officinal ditto c.m. 121 ————- exaltata Tall ditto l.b. 122 ————- tuberosa Orange Apocynum or ditto l.b. 123 ————- nigra Black ditto c.m. 124 Heuchera americana American Spanicle c.m. 125 Gentiana lutea Yellow Gentian l.b. 126 ———— saponaria Soapwort-leaved ditto l.b. 127 ————- cruciata Cross-wort ditto l.b. 128 Eryngium planum Flat-leaved Eryngo l. 129 ———— amethystinum Amethystian ditto l. 130 ———— Bourgati Cut-leaved ditto l. 131 —— —— alpinum Alpine ditto l. 132 Astrantia major Great Black Masterwort c.m. 133 Ferrula communis Gigantic Fennel l. 134 ————- nodiflora Knotted ditto l. 135 Laserpitium latifolium Broad-leaved Laser-wort l. 136 Heracleum elegans Elegant-leaved Cow Parsnep c.m. 137 Ligusticum laevisticum Common Lovage c.m. 138 ————

—— peloponnese Hemlock-leaved ditto c.m. 139 Angelica archangelica Garden Angelica c.m. 140 Sium Falcaria Creeping-rooted Skirret l.b. 141 Phellandrium Mutellina Mountain Phellandrium l.b. 142 Chaerophyllum bulbosum Bulbous-rooted Chaerophyllum c.m. 143 ——————— hirsutum Hairy ditto c.m. 144 ——————— aromaticum Sweet-scented ditto c.m. 145 Sesseli montanum Long-leaved Meadow-saxifrage c.m. 146 Thapsia villosa Deadly carrot c.m. 147 Smyrnium aureum Golden Alexanders l.b.

PENTANDRIA PENTAGYNIA.

148 Aralia racemosa Berry-bearing Aralia c.m. 149 Aralia nudicaulis Naked-stalk'd Atalia l.b. 150 Statice Cephalotes Large single-stalk'd Statice l. 151 ———- speciosa Plaintain-leaved ditto l. 152 ———- tatarica Tartarian ditto l.

HEXANDRIA MONOGYNIA.

153 Tradescantia virginica Virginian Spider-wort c.m. 154 Narcissus angustifolius Narrow-leaved Narcissus c.m. 155 ————- biflorus Two-flowered ditto c.m. 156 ————- majalis Late-flowering white ditto c.m. 157 Narcissus incomparabilis Peerless Daffodil c.m. 158 ————- major Large ditto c.m. 159 ————- orientalis Oriental ditto c.m. 160 ————- Tazetta Polyanthus Narcissus c.m. 161 ————- odorus Sweet-scented ditto c.m. 162 ————- Jonquilla Jonquil c.m. 163 ————- hispanicus Spanish-white ditto c.m. 164 ————- Bulbocodium Hoop Petticoat ditto l.b. 165 ————- minor Lesser daffodil c.m. 166 Amaryllis lutea Yellow Amaryllis l. 167 Allium victorialis Long rooted Garlick c.m. 168 ——— sphaerocephalon Small round-headed ditto c.m. 169 ——— descendens Purple-headed ditto c.m. 170 ——— nutans Nodding ditto c.m. 171 ——— senescens Narcissus-leaved

Garlick c.m. 172 ―――― multibulbosum Broad-leaved ditto c.m. 173 ―――― flavum Yellow Garlick c.m. 174 ―――― Moly Yellow Moly c.m. 175 ―――― tartaricum Tartarian Garlick c.m. 176 ―――― subhirsutum Hairy ditto c.m. 177 ―――― pallens Pale-flowered ditto c.m. 178 Lilium candidum White Lilly c.m. 179 ―――― bulbiferum Orange ditto c.m. 180 ―――― pomponium Pomponian ditto b.m. 181 ―――― chalcedonium Scarlet Martagon ditto c.m. 182 ―――― superbum Superb ditto b.m. 183 ― martagon Common Martagon ditto c.m. 184 ―――― canadense Canada-Martagon ditto b.m. 185 ―――― tigrinum Tiger Lily l.b. 186 ―――― philadelphicum Philadelphia Lily b.m.s. 187 ―――― Catesbaei Catesby's Lily b.m.s. 188 Fritillaria imperialis Crown Imperial c.m. 189 ―――――- persica Persian Fritillary l. 190 ――――――- pyrenaica Pyrenean Fritillary c.m. 191 Uvularia perfoliata Perfoliate Uvularia l.b. 192 ―――― amplexifolia Heart-leaved ditto l.b. 193 ―――― grandiflora Large-flowered ditto c.m. 194 Erythronium Dens Canis Dog's-tooth Violet c.m. 195 Tulipa sylvestris Italian Yellow Tulip c.m. 196 ―――― Gesneriana Common Garden ditto c.m. 196 Hypoxis erecta Upright Hypoxis c.m. 197 Ornithogalum nutans Nodding Star of Bethlehem c.m. 198 ―――――― pyrenaicum Pyrenean ditto c.m. 199 ――-―― latifolium Broad-leaved ditto c.m. 200 Scilla peruviana Peruvian-Hyacinth c.m. 201 ―――― campanulata Spansh Squill c.m. 202 ―――― bifolia Two-leaved ditto l.b. 203 ―――― praecox Siberian ditto l.b. 204 ―――― italica Italian ditto c.m. 205 ―――― amoena Early-flowering ditto c.m. 206 Asphodelus luteus Yellow Asphodel c.m. 207 ―――――― ramosus Branching ditto c.m. 208 Anthericum ramosum Branching Anthericum c.m. 209 ―――――― Liliago Grass-leaved ditto c.m. 210 ―――― ―――― Liliastrum St. Bruno's Lily c.m. 211 Convallaria verticillata Verticillate Solomon's Seal l. 212 ―――――- racemosa Branching ditto l. 213 ――――――- stellata Starry ditto l. 214 Hyacinthus orientalis Garden Hyacinth c.m. 215 ———— romanus Roman ditto l. 216 ——— —— cernuus Nodding ditto c.m. 217 ——— Muscaria Musk ditto c.m. 218 ———— monstrosus Feathered ditto c.m. 219 ———— comosus Purple-Grape or Tassel ditto c.m. 220 ———— botryoides Blue-Grape ditto c.m. 221 ———— racemosus Starch ditto c.m. 222 Aletris Uvaria Orange-flowered Aletris l.s. 223 Yucca gloriosa Superb Adam's Needle l.s. 224 ——- filamentosa Thready ditto c.m. 225 Hemerocallis flava Yellow Day Lily c.m. 226 ———— coerulea Blue ditto l.s. 227 ————— alba White ditto l.s. 228 —— ———— fulva Tawny ditto c.m. 229 ——————— graminea Grass-leaved ditto c.m.

HEXANDRIA TRIGYNIA.

230 Rumex Patentia Patience Dock c.m. 231 ——- italicus Italian ditto c.m. 232 ——- alpinus Alpine ditto c.m.

HEXANDRIA TETRAGYNIA.

233 Saururus cernuus Lizard's Tail c.m. 234 ———— lucidus Shining-leaved ditto c.m.

OCTANDRIA MONOGYNIA.

235 Oenothera fruticosa Shrubby Oenothera c.m. 236 Oenothera Misouriensis Misour Oenothera l.b. 237 ——— —- Fraseri Fraser's ditto l.b. 238 ———— angustifolia Narrow-leaved Shrubby ditto c.m. 239 Epilobium angustissimum Narrowest-leaved Willow-herb c.m. 240 — —- Dodonaei Dodonaeus's ditto l.b.

OCTANDRIA TRIGYNIA.

241 Polygonum divaricatum Divaricated Polygonum c.m. 242 ———— scandens Climbing ditto c.m. 243 ———— - undulatum Waved-leaved ditto c.m. 244 ———— ochreatum Spear-leaved ditto c.m. 245 ———— virginicum Virginian ditto c.m.

ENNEANDRIA TETRAGYNIA.

246 Rheum Rhaponticum Rhapontic Rhubarb c.m. 247 — —- undulatum Waved-leaved ditto c.m. 248 —— palmatum Palmated-leaved ditto c.m. 249 ——- tataricum Tartarian ditto c.m. 250 ——- hybridum Bastard ditto c.m. 251 ——- compactum Compact ditto c.m.

DECANDRIA MONOGYNIA.

252 Sophora flavescens Siberian Sophora l.b. 253 ———- alopecuroides Fox-tail ditto l.b. 254 ———— australis Blue Australian ditto l.b. 255 ———— alba White ditto l.b. 256 Cassia marilandica Maryland Cassia l. 257 Dictamnus rubra Fraxinella c.m.

DECANDRIA DIGYNIA.

258 Saxifraga crassifolia Oval-leaved Saxifrage c.m. 259 ———— cordifolia Heart-leaved ditto c.m. 260 ———— - Geum Kidney-leaved ditto c.m. 261 ———— geranoides Crane's-bill-leaved ditto c.m. 262 ———— pensylvanica Pennsylvanian ditto l.b. 263 ———— hieracifolia Hawkweed-leaved ditto c.m. 264 Gypsophila paniculata Panicled Gypsophila c.m. 265 ———— altissima Tall ditto c.m. 266 Dianthus barbatus Common Sweet William c.m. 267 ———— hybridus Mule Pink c.m. 268 ———— superbus Superb ditto c.m.

DECANDRIA TRIGYNIA.

269 Cucabulus viscosus Clammy Bladder Campion c.m. 270 ———— tataricus Tartarian ditto c.m. 271 ———— - stellatus Starry ditto l.b. 272 Silene longiflora Long-flowered Catchfly c.m.

DECANDRIA PENTAGYNIA.

273 Sedum majus Great Stonecrop c.m. 274 ——— - Aizoon Yellow ditto c.m. 275 Agrostemma coronaria Common Rose Campion c.m. 276 ———— Flos Jovis Umbell'd ditto c.m. 277 Lychnis chalcedonia Scarlet Lychnis c.m. 278 Cerastium repens Creeping Mouse-ear Chickweed c.m. 279 ———— dioicum Spanish ditto c.m. 280 ———— tomentosum Wooly-leaved ditto c.m. 281 ———— sufruticosum Shrubby ditto c.m. 282 ———— strictum Upright ditto c.m.

DECANDRIA DECAGYNIA.

283 Phytolacca decandra Branching Phytolacca l.b.

DODECANDRIA MONOGYNIA.

284 Lythrum virgatum Fine-branched Willow-herb c.m.

DODECANDRIA DIGYNIA.

285 Agrimonia odorata Sweet-scented Agrimony c.m. 286 ———— repens Creeping ditto c.m. 287 ———— Agrimonoides Three-leaved ditto c.m.

DODECANDRIA TRIGYNIA.

288 Euphorbia coralloides Coral-stalk'd Spurge l. 289 ——— —— pilosa Hairy ditto l. 290 ———— Esula Gromwell-leaved ditto l. 291 ———— falcata Sickle-leaved ditto l.

292 ———— Cyparissias Cypress ditto c.m. 293 ———— palustris Marsh ditto l.b. 294 ———— verrucosa Warted ditto l. 295 ———— multicorymbosa Flax-leaved ditto c.m.

DODECANDRIA PENTAGYNIA.

296 Spiraea Aruncus Goat's-beard Meadow Sweet c.m. 297 ———— lobata Lobe-leaved ditto l. 298 ———— trifoliata Three-leaved ditto l.b.

ICOSANDRIA POLYGYNIA.

299 Fragaria monophylla One-leaved Strawberry c.m. 300 ———— virginiana Virginian ditto c.m. 301 ———— grandiflora Pine ditto c.m. 302 ———— chiliensis Chili or White ditto c.m. 303 Potentilla pensylvanica Pensylvanian Cinquefoil c.m. 304 ———— recta Upright ditto c.m. 305 ———— hirta Hairy ditto c.m. 306 ———— mutlifida Cut-leaved ditto c.m. 307 ———— norwegica Norway ditto c.m. 308 Potentilla grandiflora Great-flowered Cinquefoil c.m. 309 ———— monspeliensis Montpelier ditto c.m. 310 Geum virginicum Virginian Avens c.m. 311 —— strictum Upright ditto c.m. 312 —— potentilloides Cinquefoil ditto c.m. 313 —— montanum Mountain ditto c.m.

POLYANDRIA MONOGYNIA.

314 Actea racemosa American Herb-Christopher c.m. 315 Podophyllum peltatum Duck's-foot, or May-apple c.m. 316 Chelidonium laciniatum Cut-leaved Celandine c.m. 317 Papaver orientale Oriental Poppy c.m.

POLYANDRIA DIGYNIA.

318 Paeonia coralloides Female Paeony l. 319 ——— humilis Dwarf ditto l. 320 ——— albiflora White-flowered ditto l. 321 ——— officinalis Common or Male ditto c.m. 322 ——— tenuiflora Fine-leaved ditto c.m. 323 ——— fimbriata Fringed-flowered ditto c.m. 324 ——— anomala Siberian ditto c.m.

POLYANDRIA TRIGYNIA.

325 Delphinium intermedium Palmate-leaved Bee Larkspur c.m. 326 ——————— hybridum Bastard ditto l. 327 ——— ——— elatum Common ditto c.m. 328 ——————— exaltatum American ditto c.m. 329 ——————— grandiflorum Large-flowered ditto c.m. 330 Aconitum Lycoctonum Great Yellow Wolf's-bane c.m. 331 ——— - Napellus Common Blue Wolf's-bane c.m. 332 ———— pyrenaicum Pyrenean ditto c.m. 333 ———— japonicum Japan ditto l.b. 334 ———— Anthora Wholesome ditto c.m. 335 ———— variegatum Variegated ditto c.m. 336 ———— ochroleucum Tall ditto c.m. 337 ———— album White-flowered ditto l. 338 ———— volubile Twining ditto l.b. 339 ———— uncinatum Hook-seeded ditto c.m. 340 ———— Cammarum Purple ditto c.m.

POLYANDRIA PENTAGYNIA.

341 Aquilegia canadensis Canadian Columbine c.m. 342 — ———- montana Mountain ditto l. 343 ———— sibirica Siberian ditto l. 344 ————- viridiflora Green-flowered ditto l.

POLYANDRIA PENTAGYNIA.

345 Anemone pratensis Meadow Anemone l.b. 346 Anemone coronaria Common Garden ditto l. 347 ——— sylvestris Snow-drop ditto c.m. 348 ——— virginiana

Virginian ditto c.m. 349 ———- pensylvanica Pensylvanian ditto c.m. 350 Clematis recta Upright Virgin's-Bower c.m. 351 ——— ochroleuca Yellow ditto l. 352 ——— viorna Leathery-flowered ditto l. 353 ——— integrifolia Intire-leaved ditto c.m. 354 Thalictrum aquilegifolium Feathered Columbine c.m. 355 ——— simplex Simple-stalked ditto c.m. 356 ——— —— lucidum Shining-leaved Meadow Rue c.m. 357 —— ——— nigricans Black-flowered ditto c.m. 358 ——— — elatum Tall ditto c.m. 359 ——— foetidum Stinking ditto c.m. 360 ——— purpurascens Purple-stalked ditto c.m. 361 ——— medium German ditto c.m. 362 ——— atropurpureum Dark-purple-flowered ditto c.m. 363 ——— rugosum Rough-leaved ditto c.m. 364 ——— dioicum Dioicous ditto c.m. 365 — ——— sibiricum Siberian ditto c.m. 366 ——— tuberosum Tubrous-rooted ditto c.m. 367 ——— angustifolium Narrow-leaved ditto c.m. 368 ——— contortum Twisted-stalked ditto c.m. 369 ——— Cornuti Canadian ditto c.m. 370 Thalictrum speciosum Glaucous-leaved Meadow Rue c.m. 371 Ranunculus aconitifolius Fair Maids of France c.m. 372 ——— platanifolius Plane-leaved Ranunculus c.m. 373 ——— — illyricus Illyrian ditto l.b. 374 ——— asiaticus Common Persian ditto c.m. 375 Trollius asiaticus Asiatic Globe-flower l.b.s. 376 ——— americanus American ditto l.b.s. 377 Helleborus niger Christmas Rose l.s. 378 — ——— lividus Livid Hellebore l.b.s.

DIDYNAMIA GYMNOSPERMA.

379 Teucrium lucidum Shining-leaved Germander c.m. 380 ——— multiflorum Many-flowered ditto c.m. 381 Hyssopus nepetoides Square-stalked Hyssop l. 382 Nepeta pannonica Hungarian Cat-Mint c.m. 383 ——— incana Hoary ditto c.m. 384 ——— violacea Violet-flowered ditto

c.m. 385 ―― Nepetella Small ditto c.m. 386 ―― nuda Spanish ditto c.m. 387 ―― tuberosa Tuberous-rooted ditto c.m. 388 Sideritis hyssopifolia Hyssop-leaved Iron-wort l. 389 ――― scordioides Crenated ditto l. 390 ――― hirsuta Hairy ditto 391 Mentha crispa Curled-leaved Mint c.m. 392 Mentha niliaca White Mint c.m. 393 ―― auriculata Ear-leaved ditto c.m. 394 Lamium Orvala Balm-leaved Archangel l. 395 ―― rugosum Wrinkled-leaved ditto c.m. 396 ―― garganicum Wolly ditto c.m. 397 ―― molle Pellitoria-leaved ditto c.m. 398 Betonica stricta Danish Betony c.m. 399 ――― incana Hoary ditto c.m. 400 ――― orientalis Oriental ditto c.m. 401 ――― hirsuta Hairy ditto c.m. 402 Stachys circinata Blunt-leaved Stachys c.m. 403 ――― lanata Woolly-leaved ditto c.m. 404 ――― cretica Cretan ditto c.m. 405 ――― recta Upright ditto c.m. 406 Marrubium supinum Procumbent Base Horehound c.m. 407 ――― hispanicum Spanish ditto c.m. 408 ― ――― peregrinum Saw-leaved ditto c.m. 409 Phlomis tuberosa Tuberous-rooted Phlomis c.m. 410 ――- Herba venti Rough-leaved ditto l.b. 411 Origanum hybridum Bastard ditto l.b. 412 ――― heracloticum Winter ditto c.m. 413 Thymus virginicus Virginian Thyme l. 414 Melissa grandiflora Great-flowered Balm c.m. 415 ――― graeca Grecian ditto c.m. 416 Dracocephalum virginicum Virginian Dragon's-head l. 417 ――― ruyschianum Hyssop-leaved ditto c.m. 418 ――― sibiricum Siberian ditto c.m. 419 Scutellaria albida Hairy Skull-cap c.m. 420 ――― integrifolia Entire-leaved ditto l.b. 421 ――― lupulina Great-flowered ditto l.b.

DIDYNAMIA ANGIOSPERMIA.

422 Chelone glabra White-flowered Chelone l.b. 423 ――― obliqua Red ditto l.b. 424 ――― ruelloides Scarlet

ditto l.b. 425 ———- formosa Tall ditto l.b. 426 Antirrhinum purpureum Purple Toad-flax c.m. 427 ———
——- genistifolium Broom-leaved ditto l. 428 ———
- triornithophorum Whorl-leaved ditto l.b. 429 Scrophularia betonicaefolia Betony-leaved Figwort l. 430 ———
orientalis Oriental ditto l. 431 Digitalis lutea Yellow Foxglove c.m. 432 ————- ambigua Great ditto c.m. 433 ————- ferruginea Iron-coloured ditto c.m. 434 Dodartia orientalis Eastern Dodartia l. 435 Penstemon pubescens American Penstemon l.b. 436 ———
laevigatum Smooth-leaved ditto l.b. 437 Mimulus ringens Oblong-leaved Monkey-flower l. 438 Mimulus guttatus Yellow Monkey-flower l.b. 439 Acanthus mollis Smooth Bear's-Breech c.m. 440 ———— spinosa Prickly ditto c.m.

TETRADYNAMIA SILICULOSA.

441 Myagrum perenne Perennial Gold-of-Pleasure c.m. 442 Cochlearia Draba Draba-leaved Scurvy-Grass c.m. 443 Iberis sempervirens Evergreen Candy-Tuft c.m. 444 Alyssum saxatile Shrubby Madwort c.m. 445 Lunaria rediviva Perennial Honesty c.m.

TETRADYNAMIA SILIQUOSA.

446 Sisymbrium strictissimum Spear-leaved Sisymbrium c.m. 447 Hesperis matronalis Single Garden Rocket c.m. 448 Bunias orientalis Oriental Bunias c.m.

MONADELPHIA DECANDRIA.

449 Geranium aconitifolium Aconite-leaved Crane's-bill c.m. 450 ———— angulosum Angular-stalked ditto c.m. 451 ———— maculatum Spotted ditto c.m. 452 ————
macorhizum Long-rooted ditto c.m. 453 ———— palustre

Marsh ditto l. 454 ———— reflexum Reflexed-flowered ditto c.m. 455 ———— striatum Striped-flowered ditto c.m. 456 ———— lividum Wrinkled ditto c.m.

MONADELPHIA POLYANDRIA.

457 Althaea cannabina Hemp-leaved Marsh-Mallow c.m. 458 Lavatera thuringiacea Large-flowered Lavatera c.m. 459 Alcea rosa Common Holyoak c.m. 460 Hibiscus palutris Marsh Hibiscus l.b. 461 Kitiabella vitifolia Vine-leaved Kitiabella c.m.

DIADELPHIA DECANDRIA.

462 Ononis antiquorum Tall Rest-Harrow l. 463 Lupinus perennis Perennial Lupine l.b. 464 Glycine Apios Tuberous-rooted Glycine l. 465 Orobus Lathyroides Upright Bitter-Vetch c.m. 466 ———— angustifolius Narrow-leaved ditto l.b. 467 ———— niger Black-flowered ditto c.m. 468 ———— vernus Spring ditto l. 469 Lathyrus tuberosus Tuberous-rooted Lathyrus c.m. 470 ———— heterophyllus Various-leaved ditto c.m. 471 ———— pisiformis Siberian ditto c.m. 472 Vicia pisiformis Pale-flowered Vetch c.m. 473 Glycyrrhiza echinata Prickly-leaved Liquorice c.m. 474 ———————- glabra Common ditto c.m. 475 Coronilla varia Purple Coronilla c.m. 476 Hedysarum canadense Canada Saintfoin c.m. 477 Galega officinalis Official Goat's-rue c.m. 478 ———— montana Mountain ditto l.b. 479 Phaca alpina Alpine Phaca, or Bastard-Vetch l.b. 480 Astralagus alopecuroides Foxtail Milk-Vetch l.b. 481 ——————- virescens Green-flowered ditto c.m. 482 ——————- galegiformis Goat's-rue-leaved ditto c.m. 483 ——————- Cicer Bladder-podded ditto l.b. 484 ——————- Onobrichis Purple-spiked ditto c.m. 485 Trifolium hybridum Bastard Trefoil, or Clover c.m. 486 — ———— rubens Long-spiked ditto c.m. 487 ————

alpestre Oval-spiked ditto c.m. 488 ——————- Lupinaster Bastard Lupine c.m. 489 Lotus maritimus Sea Bird's-foot Trefoil c.m. 490 Medicago Karstiensis Creeping-rooted Medick c.m. 491 —————— prostrata Procumbent ditto c.m.

POLYADELPHIA POLYANDRIA.

492 Hypericum calycinum Great-flowered St. John's-wort c.m.s. 493 ——————- perfoliatum Perfoliate ditto c.m.s. 494 ——————- Ascyron Red-leavedditto c.m.s.

SYNGENESIA POLYGAMIA AEQUALIS.

495 Scorzonera hispanica Spanish Viper's-grass c.m. 496 Sonchus sibiricus Siberian Sow-thistle c.m. 497 Prenanthes purpurea Purple Prenanthes l. 498 Hieracium amplexicaule Heart-leaved Hawkweed c.m. 499 ——————- pyrenaicum Pyrenean ditto c.m. 500 Crepis pontica Roman Crepis c.m. 501 Catananche caerulea Blue Catananche c.m. 502 Serratula praealta Tall Saw-wort c.m. 503 ——————- coronata Lyre-leaved ditto c.m. 504 ——————- spicata Spike-flowered ditto b.l. 505 Carduus canus Hoary Thistle c.m. 506 ————- ciliatus Ciliated ditto c.m. 507 ——————- tuberosus Tuberous-rooted ditto c.m. 508 ——————- serratuloides Saw-wort ditto c.m. 509 Cnicus oleraceus Pale-flowered Cnicus c.m. 510 ———— ferox Prickly ditto c.m. 511 ———— centauroides Centaury ditto c.m. 512 Cynara Scolymus French Artichoke c.m. 513 Carthamus corymbosus Umbelled Carthamus l.b. 514 Carline acaulis Stemless Carline l.b.s. 515 Cacalia hastata Spear-leaved Cacalia c.m. 516 ———— suaveolens Sweet-scented ditto c.m. 517 ———— saracenica Creeping-rooted ditto c.m. 518 Eupatorium maculatum Spotted Eupatorium c.m. 519 —— ———— altissimum Tall ditto c.m. 520 Eupatorium trifoliatum Three-leaved Eupatorium c.m. 521 —————— perfoliatum Perfoliate ditto l.b. 522 ——————

Ageratoides Nettle-leaved ditto b.l. 523 Chrysocoma linosyris German Goldy-locks c.m. 524 ————— biflora Two-flowered ditto c.m.

SYNGENESIA POLYGAMIA SUPERFLUA.

525 Tanacetum macrophyllum Various-leaved Tansy c.m. 526 ————- Balsamita Cost-Mary c.m. 527 Artemisia Abrotanum Common Southernwood c.m. 528 ————- santonicum Tartarian ditto or Wormseed c.m. 529 ——— —- pontica Roman ditto c.m. 530 ————— Dracunculus Tarragon c.m. 531 Conyza linifolia Flax-leaved Flea-bane c.m. 532 Tussilago paradoxa Downy-leaved Coltsfoot c.m. 533 ————- lobata Lobated ditto c.m. 534 ————- alba White ditto c.m. 535 Senecio luridus Dingy-coloured Groundsel c.m. 536 ———- coriaceus Thick-leaved ditto c.m. 537 Dahlia superflua Purple Dahlia c.m. 538 ——— v. rosea c.m. 539 ——— frustranea Red ditto c.m. 540 — ——- v. lutea Yellow ditto c.m. 541 ———- v. violacea Violet ditto c.m. 542 Boltonia asteroides Aster-leaved Boltonia c.m. 543 Aster hyysopifolius Hyssop-leaved Aster c.m. 544 ——— dumosus Purple-flowered ditto c.m. 545 — —- ericoides Heath-leaved ditto c.m. 546 ———- multiflorus Many-flowered ditto c.m. 547 ———- linearifolus Linear-leaved ditto c.m. 548 ———- foliolosus Many-leaved ditto c.m. 549 ———- salicifolius Willow-leaved ditto c.m. 550 — —- linifolius Flax-leaved ditto c.m. 551 ———- rigidus Rough-leaved ditto c.m. 552 ———- acris Biting ditto c.m. 553 ———- umbellatus Umbel'd ditto c.m. 554 ———- novae anglicae New England ditto c.m. 555 ———- grandiflorus Great-flowered ditto c.m. 556 ———- patens Spreading ditto c.m. 557 ———- aestivus Labrador ditto c.m. 558 ——— undulatus Wave-leaved ditto c.m. 559 ———- concolor Woolly ditto c.m. 560 ———- Amellus Italian ditto c.m. 561 ———- sibiricus Siberian ditto c.m. 562 ———- flexuosus Bending-stalk'd ditto c.m. 563 ———- divaricatus

Divaricated ditto c.m. 564 ——- longifolius Long-leaved ditto c.m. 565 ——- cordifolius Heart-leaved ditto c.m. 566 Aster corymbosus Purple-stalk Aster c.m. 567 ——- paniculatus Smooth-stalked panicled ditto c.m. 568 ——- puniceus Small Purple-stalked ditto c.m. 569 ——- laevis Smooth ditto c.m. 570 ——- novi belgii New-Holland ditto c.m. 571 ——- Tradescanti Tradescant's ditto c.m. 572 ——- pendulus Pendulous ditto c.m. 573 ——- diffusus Diffuse red-flowered ditto c.m. 574 ——- divergens Spreading downy-leaved ditto c.m. 575 ——- tardiflorus Spear-leaved ditto c.m. 576 ——- spectabilis Showy ditto c.m. 577 ——- mutabilis Variable ditto c.m. 578 ——- macrophyllus Broad-leaved-white ditto c.m. 579 ——- fragilis Brittle ditto c.m. 580 ——- junceus Slender-stalked ditto c.m. 581 ——- elegans Elegant ditto c.m. 582 ——- glaberrimus Smooth ditto c.m. 583 ——- lucidus Shining ditto c.m. 584 ——- sessiliflorus Sessil-flowered ditto c.m. 585 ——- altissimus Tallest ditto c.m. 586 Solidago viminea Twiggy Golden Rod c.m. 587 ——— mexicana Mexican ditto c.m. 588 ——— sempervirens Narrow-leaved Evergreen do. c.m. 589 ——— elliptica Oval-leaved ditto c.m. 590 ——— stricta Willow-leaved ditto c.m. 591 ——— latifolia Broad-leaved ditto c.m. 592 —— laevigata Fleshy-leaved ditto c.m. 593 ——— caesia Maryland ditto c.m. 594 ——— lateriflora Red-stalked ditto c.m. 595 ——— altissima Tall ditto c.m. 596 ——— arguta Sharp Notched ditto c.m. 597 ——— canadensis Canadian ditto c.m. 598 ——— procera Great ditto c.m. 599 ——— reflexa Reflexed ditto c.m. 600 ——— lanceolata Grass-leaved ditto c.m. 601 ——— serotina Upright ditto c.m. 602 ——— nemoralis Woolly-stalked ditto c.m. 603 ——— bicolor Two-cloured ditto c.m. 604 ——— aspera Rough-leaved ditto c.m. 605 ——— flexicaulis Crooked-stalked ditto c.m. 606 ——— ambigua Angular-stalked ditto c.m. 607 ——— rigida Hard-leaved ditto c.m. 608 Cineraria sibirica

Heart-leaved Cineraria c.m. 609 Inula squarrosa Net-leaved Inula c.m. 610 ——- salicina Willow-leaved ditto l.b. 611 ——- ensifolia Sword-leaved ditto c.m. 612 Helenium autumnale Smooth Helenium c.m. 613 Chrysanthemum corymbosum Large White Chrysanthemum c.m. 614 Chrysanthemum indicum Purple Indian Chrysanthemum c.m. 615 ——————- millefoliatum Tansy-leaved ditto c.m. 616 ——————- v. ——- a Quilled White. 617 — ——————- v. ——- b Double White. 618 —————— - v. ——- c Bright Yellow. 619 —————— - v. ——- d Straw-coloured 620 —————— - v. ——- e Quilled Straw-coloured. 621 —————— - v. ——- f Purple Quilled. 622 —————— - v. ——- g Lilac-coloured. 623 —————— - v. ——- h Spanish brown. 624 ——— ——— - v. ——- i Copper-coloured. 625 —————— v. ——- j Quilled Lilac. 626 Achillea alpina Alpine Millefoil or Maudlin c.m. 627 ——— cristata Slender-branched ditto c.m. 628 ——— serrata Saw'd-leaved ditto c.m. 629 ——— impatiens Impatient ditto c.m. 630 ——— santolina Lavender-Cotton-leaved ditto c.m. 631 ——— tanacetifolia Tansy-leaved ditto c.m. 632 ——— — nobilis Showy ditto c.m. 633 ——— abrotanifolia Southernwood-leaved ditto c.m. 634 Buphthalmum grandiflorum Great-flowered Ox-eye l. 635 —————- salicifolium Willow-leaved ditto l.

SYNGENESIA POLYGAMIA FRUTRANEA.

636 Helianthus multiflorus Perennial Sun-flower c.m. 637 ——————— tuberosus Jerusalem Artichoke c.m. 638 ——— ——— divaricatus Rough-leaved Sun-flower c.m. 639 — ——— decapetalus Ten-petal'd ditto c.m. 640 ——— — altissimus Tall ditto c.m. 641 ——— giganteus Gigantic ditto c.m. 642 Rudbeckia laciniata Broad-jagged-leaved Rudbeckia c.m. 643 ——————- digitata Narrow-jagged-leaved do. c.m. 644 ——— fulgida Bright

purple do. l.b. 645 ——————- purpurea Common purple do. l.b. 646 Coreopsis verticillata Whorl-leaved Coreopsis c.m. 647 —————- tripteris Three leaved ditto c.m. 648 ————- aurea Hemp-leaved ditto c.m. 649 Coreopsis procera Tall Coreopsis c.m. 650 ——————- alternifolia Alternate-leaved ditto c.m. 651 ——————- auriculata Ear-leaved ditto c.m. 652 ——————- minima Least ditto l.b. 653 Centaurea Cenaureum Great Centaury c.m. 654 ——————- alpina Alpine ditto l.b. 655 ——————- montana Mountain ditto c.m. 656 ——————- sempervirens Evergreen ditto c.m. 657 ——————- sibirica Siberian ditto c.m. 658 ——————- phrygia Austrian ditto c.m. 659 Centaurea glastifolia Woad-leaved Centaury l.b. 661 ——————- rhapontica Swiss ditto l.b. 662 ——————- sonchifolia Sow-thistle-leaved ditto l.b. 663 ——————- aurea Great Yellow ditto l.b.

SYNGENESIA POLYGAMIA NECESSARIA.

664 Silphium scabrum Rough-leaved Silphium c.m. 665 —————- terebinthinum Broad-leaved ditto c.m. 666 ————— perfoliatum Perfoliate ditto c.m. 667 ————— connatum Round-stalked ditto c.m. 668 ————— Asteriscus Hairy-stalked ditto c.m. 669 ————— trifoliatum Three-leaved ditto c.m.

SYNGENESIA POLYGAMIA SEGREGATA.

670 Echinops Ritro Small Globe Thistle c.m. 671 ——————— sphaerocephalus Great ditto c.m.

SYNGENESIA MONOGAMIA.

672 Lobelia Cardinalis Scarlet Cardinal flower l. 673 ——————- siphylitica Blue ditto l.

GYNANDRIA TRIANDRIA.

674 Sisyrinchium striatum Striated Sisyrinchium l.

GYNANDRIA POLYANDRIA.

675 Arum Dracunculus Long-sheathed Arum c.m. 676 —— venosum Varied ditto c.m.

MONOECIA PENTANDRIA.

677 Parthenium integrifolium Intire-leaved Parthenium c.m. 678 Urtica nivea Snowy Nettle c.m.

DIOECIA HEXANDRIA.

669 Smilax herbacea Herbaceous Smilax b.l.s.

DIOECIA DODECANDRIA.

680 Datisca cannabina Bastard Hemp c.m.

DIOECIA MONADELPHIA.

681 Napaea laevis Smooth Napaea l.b. 682 ——- scabra Rough ditto c.m.

POLYGAMIA MONOECIA.

683 Veratrum album White Hellebore l.b.s. 684 ——— nigrum Dark-flowered Veratrum l.b.s.

SECTION XVIII.-HARDY ANNUAL FLOWERS.

These are cultivated by sowing their seeds, in the months of March or April, in the places where they are to remain and flower during the summer months.

ENGLISH NAMES. LATIN NAMES.

1 Alyssum sweet Alyssum halimifolium 2 Alkekengi Physalis Alkakengi 3 Arctotus annual Arctotus anthemoides 4 Argemone or Devil's Fig Argemone mexicana 5 Asphodel annual Anthericum anuum 6 Aster China quilled 7 ——- red Aster chinensis 8 ——- white Aster chinensis 9 ——- purple Aster chinensis 10 —— superb Aster chinensis 11 —— Bonnet Aster chinensis 12 —— striped Aster chinensis 13 Balm Moldavian Dracocephalon moldavicum 14 —— white Dracocephalon moldavicum 15 —— hoary Dracocephalon moldavicum 16 Belvidera Chenopodium Scoparium 17 Bladder Ketmia Hibiscus trionum 18 Candytuft purple Iberis umbellata 19 ————- white Iberis umbellata 20 ————- Normandy Iberis umbellata 21 Caterpillar Scorpiurus vermiculata 22 Catchfly pendulous Silene pendula 23 ———— Lobel's Armeria 24 Cyanus major Centaurea Crupina 25 ——— minor Centaurea Cyanus 26 Clary purple topped Salvia Hormium 27 ——- Red ditto Salvia Hormium 28 Chrysamthemum white-quill'd Chrysamthemum coronarium 29 ————- yellow ditto Chrysamthemum tricolor 30 Hawkweed red Crepis rubra 31 ———— yellow Crepis barbata 32 Hedgehogs Medicago polymorpha, v. intertexta 33 Honeywort great Cerinthe major 34 ————- small Cerinthe minor 35 Indian Corn Zea mays 36 Jacobaea Senecio elegans 37 Larkspur Tall Rocket Delphinium Ajacis 38 ———— Dwarf Rocket

Delphinium Ajacis 39 ——— Rose Larkspur Delphinium Ajacis 40 ——— Branching ditto Delphinium Ajacis 41 Lavatera Red Lavatera trimestris 42 ——— white Lavatera trimestris 43 Lobel's Catchfly red Silene armeria 44 ——————— white Silene armeria 45 Love-lies-bleeding Amaranthus caudatus 46 Lupine yellow Lupinus luteus 47 ——— straw-coloured Lupinus luteus 48 ——— large blue Lupinus hirsutus 49 ——— small ditto Lupinus varius 50 ——— rose Lupinus pilosus 51 ——— blue Dutch Lupinus var 52 ——— white Lupinus albus 53 Mallow-curled Malva crispa 54 Marigold French Tagetes patula 55 ——— African Tagetes erecta 56 ——— small cape Calendula pluvialis 57 ——— great Cape Calendula hybrida 58 ——— starry Calendula stellata 59 Mignionette Reseda odorata 60 Nasturtium great Tropaeolum majus 61 ——— small Tropaeolum minus 62 Nettle Roman Urtica pilulifera 63 Nigella Roman Nigella Romana 64 ———- Spanish Nigella Hispanica 65 ———- small Nigella sativa 66 Nolana Trailing Noalan prostrata 67 Noli-me-Tangere Impatiens Noli-me-Tangere 68 Oenothera purple Oenothera purpurea 69 Pea sweet purple Lathyrus odoratus 70 ———- scarlet Lathyrus odoratus 71 ———- white Lathyrus odoratus 72 ———- black Lathyrus odoratus 73 ———- striped Lathyrus odoratus 74 ——— ——- painted lady Lathyrus odoratus 75 Pea jointed-podded Lathyrus articulatus 76 —- Anson's Lathyrus magellanicus 77 —- Painted Lady Crown Lathyrus sativus 78 —- Tangier scarlet Lathyrus tingitanus 79 —- purple Lathyrus tingitanus 80 —- red-winged Lotus tetragonolobus 81 —- yellow ditto Lotus tetragonolobus 82 Persicaria red Polygonum orientale 83 ——— white Polygonum orientale 84 Poppy carnation Papaver somniferum 85 ———- dwarf Rhoeas 86 Quaking-grass Briza maxima 87 Saltwort Rose Salsola rosacea 88 Scabious starry Scabiosa stellata 89 Snails Medicago scutella 90 Soapwort Saponaria Vaccaria 91 Stock purple 10-week Cheiranthus annuus 92 ——- scarlet 10-week Cheiranthus annuus 93 ——- white 10-week Cheiranthus annuus 94 ——- white Prussian Cheiranthus annuus 95 ——- purple ditto Cheiranthus annuus 96 Stock Virginian white Cheiranthus maritimus 97 ——————- red Cheiranthus annuus 98 Stramonium purple Datula Tatula 99 ———— white Datula stramonium 100 Spinage strawberry Blitum virgatum 101 Sunflower tall Helianthus annuus 102 ————- dwarf Helianthus annuus 103 ——— ——- double Helianthus annuus 104 Sultan sweet purple Centaurea moschata 105 ——— white Centaurea moschata 106 ——— yellow Centaurea suaveolens 107 Toadflax three-leaved Antirrhinium triphyllum 108 Trefoil crimson Trifolium incarnatum 109 Venus's Looking-glass Campanula speculum 110 ——-Navelwort Cynoglossum linifolium 111 Xeranthemum yellow shining Xeranthemum lucidum 112 —————- white Xeranthemum annuum 113 —————- purple double Xeranthemum annuum 114 Zinnia yellow Zinnia pauciflora 115 ——— red Zinnia multiflora 116 ——— elegant Zinnia elegans 117 ——— violet-coloured Zinnia tenniflora 118 ——— whorl-leaved Zinnia verticillata

SECTION XIX.-BIENNIAL FLOWERS.

Biennial Flowers, i.e. such as do not bloom the same year they are raised from seeds.

These should be sown in the month of May or June, and let remain in the place till the month of September, when they should be planted into beds, and in the following spring placed out where they are to flower.

1 Canterbury Bells Campanula media 2 Iron-coloured Foxglove Digitalis ferruginea 3 Hollyoak Alcea rosa 4 Honesty Lunaria rediviva 5 Stocks red Brompton Cheiranthus incanus 6 ——— white ditto Cheiranthus incanus 7 ——— purple ditto Cheiranthus incanus 8 ——— Queen Cheiranthus incanus 9 ——— Twickenham Cheiranthus incanus 10 Wallflower Cheiranthus fruticulosus

SECTION XX.-TENDER ANNUAL FLOWERS.

Such as are usually sown in hot-beds in the months of February or March, and grown in the stove or green-house after the removal of the plants in the summer months, for which purpose they are very ornamental.

ENGLISH NAMES LATIN NAMES

1 Amaranthus three-coloured Amaranthus tricolor 2 ——— ——— two-coloured ——— bicolor 3 ——— globe white Gomphrena globosa 4 ——— purple Gomphrena globosa 5 Balsam Impatiens Balsamita 6 ——— — scarlet Impatiens coccinea 7 Striped double white 8 Browallia blue Browallia elata 9 ———- white Browallia elata 10 Cacalia scarlet Cacalia coccinea 11 Capsicum large red Capsicum annuum 12 ——— yellow Capsicum annuum 13 ——— small red horn Capsicum annuum 14 ——— yellow ditto Capsicum annuum 15 — ——— cherry Capsicum annuum 16 ——— Cayenne Capsicum annuum 17 Calceolaria wing-leaved Calceolaria pinnata 18 Convolvulus large-flowered Convolvulus major 19 ———- minor ———- tricolor 20 Cockscomb dwarf Celosia cristata 21 ———- tall Celosia cristata 22 ———- branching Celosia cristata 23 ———- buff or yellow Celosia cristata 24 Egg plant white Solanum Melongena 25 ———- purple Solanum Melongena 26 Impomaea Scarlet Impomaea coccinea 27 — ——- wing-leaved ———- Quamoclit 28 Ice plant Mesembryanthemum crystallinum 29 Love apple Solanum Lycopersicum 30 Sensitive plant Mimosa pudica 31 Stramonium double purple Datura Metel 32 ——— Double white ——— v. flore albo

SECTION XXI.-FOREIGN ALPINE PLANTS.

ADAPTED TO THE DECORATION OF ROCK-WORK.

The following list comprises a number of plants of great beauty and interest; but, being in general too small for the open borders, are only to be preserved either in pots; planted in rock-work, or in such other places where they are not overgrown by plants of larger size. There are many others of a similar kind that we grow in gardens, but which, being difficult to keep, we have thought fit not to insert; as persons who try to cultivate such in the open ground have in general the mortification to find that they do not compensate for the care and trouble necessary for preserving them.

1 Ancistrum lucidum Shining Ancistrum b.l. 2 ———— laevigatum Smooth ditto b.l. 3 ———— latebrosum Hairy ditto b.l. 4 Veronica aphylla Naked-stalked Speedwell b.l. 5 ———— bellidoides Daisy-leaved ditto b.l.

TRIANDRIA MONOGYNIA.

6 Trichonema Bulbocodium Crocus-leaved Trichonema b.l.

TETRANDRIA MONOGYNIA.

7 Asperula crassifolia Thick-leaved Woodroofe b.l. 8 Houstonia caerulea Blue Houstonia l. 9 Mitchella repens Creeping Mitchella l. 10 Plantago alpina Alpine Plantain l. 11 ———— subulata Awl-leaved ditto l. 12 Cornus canadensis Herbaceous Dog-wood b. 13 Alchemilla pentaphylla Five-leaved Lady's Mantle b.l. 14 ———— argentata Silvery-leaved ditto b.l.

PENTANDRIA MONOGYNIA.

15 Cynoglossum Omphaloides Blue Venus's Navelwort b.l. 16 Aretia vitaliana Primrose aretia l. 17 Androsace villosa Hairy Androsace l. 18 Primula cortusoides Bear's-ear Primrose b.l. 19 ——— villosa Hairy Primula b.l. 20 ——— — nivea Snowy ditto b.l. 21 ——— marginata Margined ditto b.l. 22 ——— Auricula Common Yellow Auricula b.l. 23 ——— lonigfolia Long-leaved ditto b.l. 24 ——— helvetica Swiss ditto b.l. 25 Primula integrifolia Entire-leaved Auricula b.l. 26 Cortusa Mathioli Siberian Bear's-ear Sanicle b. 27 Soldanella alpina Alpine Soldanella b.l. 28 Dodecatheon Meadia American Cowslip b.l. 29 Cyclamen Coum Round-leaved Cyclamen l. 30 ——— hederaefolium Ivy-leaved ditto l. 31 Lysimachia dubia Purple Loosestrife l. 32 Phlox pilosa Hairy Lychnidea l. 33 ——- ovata Oval-leaved ditto l. 34 ——- suffruticosa Shrubby ditto l. 35 ——- stolonifera Creeping ditto l. 36 — —- subulata Awl-leaved ditto l. 37 ——- setacea Bristly ditto l. 38 Convulvulus lineatus Dwarf Bindweed l. 39 Campanulla pulla Dark-flowered Bell-flower b.l. 40 ——— ——— carpatica Carpasian ditto b.l. 41 ——— pumila Purple-dwarf ditto b.l. 42 ——— v. alba White-dwarf ditto b.l. 43 ——— nitida Shining-leaved ditto b.l. 44 ——— barbata Bearded ditto b.l. 45 ——— azurea Azure-coloured ditto b.l. 46 Phyteuma hemisphaerica Small Rampion b.l. 47 Verbascum Myconi Borage-leaved Mullein l.

PENTANDRIA DIGYNIA.

48 Gentiana acaulis Gentianella l. 49 ——— asclepiadea Swallow-wort Gentian l. 50 Bupleurum petraeum Rock Thorough-wax l.

PENTANDRIA TRIGYNIA.

51 Telephium Imperati True Orphine l.

PENTANDRIA PENTAGYNIA.

52 Statice cordata Heart-leaved Thrift l.
 53 ———- flexuosa Zigzag ditto l.
 54 Linum flavum Yellow Flax l.
 55 ——- austriacum Austrian ditto l.

HEXANDRIA MONOGYNIA.

56 Convallaria bifolia Two-leaved Lilly of the Valley l.b.

HEXANDRIA TRIGYNIA.

57 Trillium cernuum Drooping-flowered Trillium b.
 58 ———— sessile Sessile-flowered ditto b.
 59 Helonias bullata Spear-leaved Helonias b.
 60 ———— asphodeloides Grass-leaved ditto b.

OCTANDRIA MONOGYNIA.

61 Rhexia mariana Hairy Rexia b. 62 Oenothera rosea Rose-flowered Tree Primrose l.b. 63 ————- pumila Dwarf Yellow ditto l.b. 64 Epilobium cordifolium Heart-leaved Willow-herb b.l.

OCTANDRIA DIGYNIA.

65 Moehringia muscosa Mossy Moehringia l.

DECANDRIA DIGYNIA.

66 Saxifraga Cotyledon Pyramidal Saxifrage l. 67 ———
—- Aizoon Margined ditto c.m. 68 ————- ligulata Strap-leaved ditto c.m. 69 ————- rosularis Rose-leaved ditto c.m. 70 ————- mutata House-leek ditto c.m. 71 —
————- Androsace Blunt-leaved ditto c.m. 72 ————-
caesia Gray ditto c.m. 73 ————- pilosa Hairy ditto c.m. 74 ————- sarmentosa Creping ditto c.m. 75 ————-
cuneifolia Wedge-leaved ditto c.m. 76 ————- aspera Rough-leaved ditto c.m. 77 ————- rotundifolia Round-leaved ditto c.m. 78 ————- ajugaefolia Ground Pine-leaved ditto c.m. 79 ————- sibirica Siberian Pine-leaved ditto c.m. 80 ————- adscendens Ascending Saxifrage c.m. 81 ————- viscosa Clammy ditto c.m. 82 Tiarella cordifolia Heart-leaved Tiarella c.m. 83 Mitella diphylla Two-leaved Mitella c.m. 84 Gypsophila repens Creeping Gypsophila l.b. 85 ————— prostrata Trailing ditto l.b. 86 Saponaria acymoides Basil-leaved Soap-wort l. 87 ———— superbus Feathered ditto l. 88 ————
pungens Pungent ditto l. 89 ———— alpinus Alpine ditto l. 90 ———— capitatus Headed-flowered ditto l. 91 ——
—— glaucus Glaucous ditto l. 92 ———— virgineus Maiden ditto l.

DECANDRIA TRIGYNIA.

93 Silene anemoena Siberian Catchfly l.
  94 ——— alpestris Mountain ditto l.
  95 ——— rupestris Rock ditto l.
  96 ——— saxifraga Saxifrage ditto l.
  97 ——— vallesia Downy ditto l.
  98 Stellaria scapigera Naked-stalk'd Stitch-wort l.
  99 Arenaria tetraquetra Square Sand-wort l.
  100 ——- balearica Small ditto l.
  101 ———- saxatilis Rock ditto l.
  102 ———- striata Striated ditto l.
  103 ———- grandiflora Great-flowered ditto l.

104 ———- liniflora Flax-flowered ditto l.

DECANDRIA PENTAGYNIA.

105 Sedum Aizoon Yellow Stonecrop c.m. 106 ——- Anacampseros Evergreen Orpine c.m. 107 ——- hybridum Bastard Sedum c.m. 108 ——- populifolium Poplar-leaved ditto c.m. 109 ——- virens Green ditto c.m. 110 ——- glaucum Glaucous ditto c.m. 111 ——- deficiens Round-leaved ditto c.m. 112 ——- hispanicum Spanish ditto l. 113 Lychnis quadridentata Small-flowering Lychnis l.b.

DODECANDRIA MONOGYNIA.

114 Asarum canadense Canadian Asarabaca l.b.

DODECANDRIA DIGYNIA.

115 Sempervivum globiferum Globular House-leek l.
116 ——————- arachnoideum Cobweb ditto l.
117 ——————- hirtum Hairy ditto l.
118 ——————- montanum Mountain ditto l.
119 ——————- cuspidatum Prickly-leaved ditto l.
120 ——————- sediforme Stone-crop-leaved ditto l.

ICOSANDRIA POLYGYNIA.

121 Rubus arcticus Dwarf Bramble l.b. 122 Potentilla sericea Silky Cinquefoil l.b. 123 —————— multifida Multifid ditto l. 124 —————— bifurca Bifid ditto l. 125 —————— tridentata Trifid-leaved ditto l. 126 Geum potentilloides Cinquefoil Avens l. 127 —— reptans Creeping ditto l.

POLYANDRIA MONOGYNIA.

128 Sanguinaria canadensis Canada Puccoon l.b. 129 Papaver nudicaule Naked-stalked Poppy l. 130 Cistus grandiflorus Great-flowered Cistus l.

POLYANDRIA POLYGYNIA.

131 Anemone Hepatica Common Liverwort c.m. 132 ——— -- hortensis Star Anemone l.b. 133 ———- dichotoma Forked ditto l.b. 134 Adonis vernalis Spring Adonis Flower c.m. 135 Ranunculus amplexicaulus Plaintain-leaved Crow-foot l.b. 136 ——————— alpestris Alpine ditto l.b. 137 ——————— glacialis Two-flowered ditto l.b. 138 Isopyrum thalictroides Thalictrum-leaved Isopyrum c.m.

DIDYNAMIA GYMNOSPERMA.

139 Teucrium multiflorum Many-flowered Germander c.m. 140 ——————— pyrenaicum Pyrenean ditto c.m. 141 Dracocephalum denticulatum Tooth-leaved Dragon's-head c.m. 142 ———————- austriacum Austrian ditto b.l. 143 ———————- grandiflorum Great-flowered ditto l. 144 Scutellaria alpina Alpine Skull-cap l. 145 ———————- grandiflora Large-flowered ditto l. 146 Prunella laciniata Cut-leaved Self-heal c.m. 147 ————— grandiflora Large-flowered ditto c.m. 148 ————— hyssopifolia Hyssop-leaved ditto c.m. 149 ————— latifolia Broad-leaved ditto c.m.

DIDYNAMIA ANGIOSPERMA.

150 Erinus alpinus Alpine Erinus l.b.

TETRADYNAMIA SILICULOSA.

151 Draba aizoides Hairy-leaved Willow-grass l.b. 152 Lepidium alpinum Mountain Pepper-wort l.b. 153 Iberis

saxatilis Rock Candy-tuft l.b. 154 Alyssum montanum Mountain Mad-wort l. 155 ——— utriculatum Bladder-podded ditto l. 156 ——— deltoideum Purple-flowered ditto l. 157 ——— campestre Small yellow ditto l.

TETRADYNAMIA SILIQUOSA.

158 Cardamine asarifolia Heart-leaved Lady's Smock l. 159 ——— bellidifolia Daisy-leaved ditto l. 160 ——— trifolia Three-leaved ditto l.b. 161 Cheiranthus alpinus Alpine Stock l. 162 Arabis alpina Alpine Wall-Cress l. 163 ——— lucida Shining-leaved ditto l. 164 ——— bellidifolia Daisy-leaved ditto l. 165 ——— sibirica Siberian ditto l.b.

MONADELPHIA PENTANDRIA.

166 Erodium Reichardi Dwarf Erodium c.m.

DIADELPHIA HEXANDRIA.

167 Fumaria cucullaria Naked-stalked Fumitory l.
168 ——— nobilis Great-flowered ditto l.
169 Fumaria cava Hollow-rooted Fumitory l.
170 ——— solida Solid-rooted ditto l.
171 ——— spectabilis Scarlet ditto l.

DIADELPHIA DECANDRIA.

172 Hedysarum obscurum Creeping-rooted Hedysarum l.b. 173 Astragalus pilosus Hairy Milk-Vetch l. 174 ——— falcatus Sickle-podded ditto l. 175 ——— uliginosus Marsh ditto l. 176 ——— monspessulanus Montpelier ditto l. 177 ——— exscapus Stalkless ditto l. 178 ——— campestris Field ditto l.

SYNGENESIA POLYGAMIA AEQUAIS.

179 Leontodon aureum Golden Dandelion l.

POLYGAMIA SUPERFLUA.

180 Artemisia glacialis Creeping Wormwood c.m. 181 Gnaphalium plantagineum Plaintain-leaved Everlasting l. 182 Erigeron philadelphicum Philadelphia Erigeron l. 183 ——— purpureum Purple ditto l.b.

SYNGENESIA MONOGAMIA.

184 Lobelia minuta Least Cardinal Flower 185 Viola palmata Palmated Violet b. 186 ——- cucullata Hollow-leaved ditto l. 187 ——- canadensis Canadian ditto l.b. 188 ——- striata Striated ditto l.b. 189 ——- pubescens Downy ditto l.b. 190 ——- biflora Two-flowered ditto l.b. 191 ——- grandiflora Great-flowered ditto l.b. 192 ——- calcarata Alpine ditto l.b. 193 ——- cornuta Pyrenean ditto l.b. 194 ——- obliqua Oblique-leaved ditto l.b. 195 Tussilago alpina Alpine Colt's-foot c.m. 196 Senecio abrotanifolia Southernwood-leaved Grounsel c.m. 197 Aster alpinus Alpine Star-wort l.b. 198 Doronicum bellidiastrum Daisy-leaved Leopard's-Bane l.b. 199 Bellis lusitania Portugal Daisy l.b. 200 Bellium minutum Bastard Daisy l.b. 201 Anthemis Pyrethrum Pellitory of Spain l.b. 202 Achillea tomentosa Woolly Milfoil l.b. 203 ——— Clavannae Silvery-leaved ditto l.b.

GYNANDRIA DIANDRIA.

204 Cypripedium album White Ladies-Slipper b.

GYNANDRIA TRIANDRIA.

205 Sisyrinchum anceps Small Sisyrinchum c.m. 206 Arum tenuifolium Fine-leaved Arum c.m.

CRYPTOGAMIA FILICES.

207 Polypodium marginale Margin-flowered Polypody b.l. 208 ———— auriculatum Eared ditto b.l. 209 Onoclea sensibilis Sensitive Fern b. 210 Equisetum filiforme Fine Horse-tail l.

APPENDIX

BRITISH PLANTS CULTIVATED FOR ORNAMENTAL PURPOSES.

1. ALISMA Plantago. I cannot pass over this beautiful aquatic without giving it a place amongst the ornamental plants with which our country abounds. In pieces of water this is of considerable interest both as to flowers and foliage, and no place of the kind should ever be destitute of such a beauty. It is of easy culture; the plant taken from its place of growth and sunk into the water with a stone to keep it in its place, is a ready and easy mode of planting it, and there is no fear when once introduced but it will succeed.

2. ANDROMEDA polifolia. This is a beautiful little shrub, and grown in gardens for the sake of its flowers; it is also an evergreen. This plant will not succeed unless it is planted in bog earth,—for a description of which see page 152 of this volume.

3. AQUILEGIA vulgaris. COLUMBINE.—We have scarcely a plant affording more beauty or greater variety than this. It is commonly, when found wild, of a blue colour, but when the seeds are sown in the garden a variety of tints is produced. It is a perennial, but easily raised from seed, which should be sown in the spring.

4. ANTHEMIS maritima. A double-flowering variety of this plant used to be common in the gardens near London, but is now scarce: it is very beautiful, and constantly in bloom during summer. It is propagated by planting the roots in the spring and autumn.

5. ANTIRRHINUM linaria, v. Peloria.—I cannot pass over this singular and beautiful flower without notice. There is a fine figure of it in the Flora Londinensis: it is very ornamental, and the structure of the bloom is truly interesting. It is easily propagated by planting the roots in the spring months, but it is not common.

6. ANTIRRHINUM majus. SNAPDRAGON.—This is also a plant deserving the attention of the lover of flowers: it is capable of culture into many very beautiful and interesting varieties.

7. BELLIS perennis. DAISY.—This plant affords us many very beautiful varieties for the flower garden. The large Red Daisy and all the other fine kinds are only this plant improved by culture.

8. BUTOMIS umbellatus. This is an aquatic, and well adapted to ornament pieces of water. Its beautiful flowers in the summer months are inferior to scarcely any plants growing in such places, and its foliage will form protection for any birds, &c., which are usually kept in such places. It is easily propagated by planting it in such places.

9. CALTHA palustris. MARSH MARIGOLD.—This fine yellow flower is also made double by culture, and finds a place in the flower garden.

10. CHEIRANTHUS fruticulosus. WALLFLOWER.—Is a plant possessing great beauty, and very interesting on account of its fine scent. We have this plant also improved by culture, making many fine double varieties. It is a biennial, and easily raised from seeds, which should be sown in June. The double varieties are cultivated by cuttings of the branches.

11. CYPRIPEDIUM Calceolus. LADIES SLIPPER.—A flower of the most uncommon beauty, but is now become scarce; it is a native of the woods near Skipton in Yorkshire, but has been so much sought for by the lovers of plants as to become almost extinct. It is difficult to propagate; but when the plants have been for some years growing, will admit of being parted, so that it may be increased in that way: it will not bear to be often removed, and should be left to grow in the same place for several years without being disturbed. It succeeds best in bog earth or rotten leaves.

12. DELPHINIUM Ajacis. LARKSPUR.—This is also an annual flower, affording a pleasing variety in the flower garden in the summer months. For it culture, see p. 188.

13. DIANTHUS Caryophyllus. THE CARNATION.—All our fine varieties of the carnation are the produce of this plant.

The common single variety produces seed in great abundance, but the improved double varieties are sparing in produce: the fine kinds of this flower are reared by layers put down about the month of July; they may also be propagated by cuttings, but the other is the most eligible and certain mode.

14. EPILOBIUM angustifolium. A plant of singular ornament. There is also a white variety of this found in gardens.

15. ERICA vulgaris. There is now in cultivation in the gardens a double-flowering variety of this plant, which is highly interesting and of singular beauty. It grows readily in bog earth, and is raised by layers.

16. ERICA Daboeica. IRISH HEATH.—A plant of singular beauty and of easy culture; and being of small growth and almost constantly in bloom, has also obtained a place in the shrubbery.

17. FRITILLARIA Meleagris. A very ornamental bulbous plant, of which the Dutch gardeners have many improved varieties, varying in the colour and size of the blossoms: these are usually imported in August, and should be immediately planted, as the bulbs will not keep long when out of ground, unless they are covered with sand.

18. GALANTHUS nivalis. SNOWDROP.—The first of the productions of Flora which reminds us of the return of spring after the dark and dreary days of winter. This plant is also made double by cultivation, but is not handsomer than the common wild one. The best time for planting the bulbs of Snowdrops is in the month of September.

19. GENTIANA verna. VERNAL GENTIAN.—A delightful little plant of the finest blue colour the Flora exhibits in all her glory: its scent is also delightful: it is somewhat scarce and difficult to procure; but if more generally known, few gardens would be destitute of such a treasure. It is of tolerably easy culture, and grows well in loam: it is small, and is best kept in a pot.

20. GENTIANA Pneumonanthe. MARSH GENTIAN.—Is also a beautiful plant, and grows well in any moist place. From its beautiful blue flowers it is well adapted to the flower garden; it delights in bog earth.

21. GERANIUM phaeum. BLACK-FLOWERED GERANIUM.—This is a perennial, and makes a fine ornamental plant for the shrubbery: it will grow in any soil and situation.

22. GLAUCUM Phoeniceum. PURPLE HORN POPPY.—An annual flower of singular beauty, and deserving a place in the flower garden.

23. GNAPHALIUM margaritaceum. AMERICAN CUDWEED.—This plant affords beautiful white flowers, which drying and keeping their colour, it is worth attention on that account, as it affords a pleasing variety with the different Xeranthema, and others of the like class in winter.

24. HIERACUM aurantiacum. GRIM-THE-COLLIER.—This is an old inhabitant of our gardens, and affords a pleasing variety.

25. HOTTONIA palustris. WATER VIOLET.—This is a plant of singular beauty in spring; it is an aquatic, and makes a fine appearance in our ponds in the time of its bloom.

26. IBERIS amara. CANDYTUFT.—An annual flower of considerable beauty and interest. We have several varieties of this sold in the seed-shops.

27. IMPATIENS NOLI ME TANGERE.—A very curious flower which is grown as an annual. The construction of the seed-vessel causing the seeds to be discharged with an elastic force is a pleasing phaenomenon.

28. LATHYRUS sylvestris.—EVERLASTING PEA.—This is also a great ornament, and frequently found in gardens; it grows very readily from seeds sown in the spring of the year.

29. LEUCOJUM aestivum. SUMMER SNOW FLAKE.—This is a very noxious plant in the meadows where it grows wild. I have seen it in the neighbourhood of Wooking in

Surrey quite overpower the grass with its herbage in the spring, and no kind of that animal that we know of will eat it.

It is however considered an ornamental plant, and is often found in our flower gardens. It is of easy culture: the roots may be planted in any of the autumn or winter months.

30. MALVA moschata. MUSK MALLOW.—This makes a fine appearance when in bloom, for which purpose it is often propagated in gardens: its scent, which is strong of vegetable musk, is also very pleasant.

31. MELLITIS mellyssophyllum. MELLITIS grandiflora. BASTARD BALM.—Both these plants are very beautiful, and are deserving a place in the flower garden: they are of easy culture, and will grow well under the shade of trees, a property that will always recommend them to the notice of the curious.

32. MENYANTHES Nymphoides. ROUND-LEAVED BOG BEAN.—This is a beautiful aquatic, and claims a place in all ornamental pieces of water.

33. NARCISSUS poeticus. NARCISSUS Pseudo Narcissus.—These are much cultivated in gardens for the sake of the flowers. The florists have by culture made several varieties, as Double blossoms which are great ornaments. The season for planting the bulbs of Narcissus of all kinds is the month of October: they will grow well in any soil, and thrive best under the shade of trees.

34. NUPHAR minima is also beautiful, but it is not common. It will form an ornament for pieces of water.

35. NYMPHAEA alba. NYMPHAEA lutea.—These are aquatics, and scarcely any plant is more deserving of our attention. The fine appearance of the foliage floating on the surface, which is interspersed with beautiful flowers, will render any piece of water very interesting: it should also be observed that gold-fish are found to thrive best when they have the advantage of the shade of these plants. It is difficult in deep water to make them take root, being liable to float on the surface, in which state they will not succeed. But if the plants are placed in some strong clay or loam tied down in wicker baskets and then placed in the water, there is no fear of their success: they should be placed where the water is sufficiently deep to inundate the roots two feet or a little more.

36. OPHRYS apifera. BEE ORCHIS.—There are few plants that are more generally admired than all the Orchideae for their singular beauty and uncommon structure. The one in question so very much resembles the humble-bee in appearance, that I have known persons mistake this flower for the animal. It is unfortunate for the amateurs of gardening that most plants of this tribe are difficult of propagation, and are not of easy culture. I have sometimes succeeded with this and other species, by the following method:—to take up the roots from their native places of growth as early as they can be found, and then procure some chalk and sift it through a fine sieve, and also some good tenacious loam; mix both in equal quantities in water; a large garden-pot should then be filled with some rubble of chalk, about one third deep, and then the above compost over it, placing the roots in the centre, at the usual depth they grew before. As the water drains away, the loam and chalk will become fixed closely round the bulbs, and they will remain alive and grow. By this method I have cultivated these plants for some years together.

In this way all those kinds growing in chalk may be made to grow; but such as the Orchis moryo, maculata, and pyramidalis, may be grown in loam alone, planted in pots in the common way. Care should be taken that the pots in which they are planted are protected from wet and frost in the winter season.

37. ORNITHOGALUM latifolium and umbellatum are also ornamental, and are often cultivated for their beautiful flower. The season for planting the bulbs is about the month of September.

38. PAPAVER somniferum. GREATER POPPY. PAPAVER Rhoeas. CARNATION POPPY. —These are made by culture into numerous varieties, and are very beautiful; but the aroma, which is pregnant with opium, renders too many of them unpleasant for the garden.

39. POLEMONIUM coeruleum. GREEK VALERIAN, or JACOB'S LADDER.—Is also a beautiful perennial, and claims the notice of the gardener. Its variety, with white flowers, is also ornamental. It is raised from seeds, which are sold in plenty in our seed-shops.

40. PRIMULA officinalis. COWSLIP. PRIMULA vulgaris. PRIMROSE. PRIMULA elatior. OXLIP. PRIMULA farinose. BIRD'S EYE.—All well known ornaments of numerous varieties, double and single. The third species is the parent of the celebrated Polyanthus. The last is also an interesting little plant with a purple flower. It grows best in bog earth.

41. ROSA rubiginosa. SWEET BRIAR.—This lovely and highly extolled shrub has long claimed a place in our gardens. We have several varieties with double flowers, which are highly prized by the amateurs of gardening.

42. SAXIFRAGA umbrosa. LONDON PRIDE.—-A beautiful little plant for forming edgings to the flower garden, or for decorating rock-work.

43. SAXIFRAGA oppositifolia. PURPLE SAXIFRAGE.— Perhaps we have few flowers early in the spring that deserve more attention than this. It blooms in the months of February and March, and in that dreary season, in company with the Snow-drop, Crocus, and Hepaticas, will form a most delightful group of Flora's rich production. The Saxifrage is a native of high mountains, and it can only be propagated by being continually exposed to the open and bleakest part of the garden: it succeeds best in pots. It should be parted every spring, and a small piece about the size of a shilling planted in the centre of a small pot, and it will fill the surface by the autumn. The soil bestsuited to it is loam.

44. SEDUM acre. STONE CROP. SEDUM rupestre. ROCK GINGER.—All the species of Sedums are very ornamental plants, and are useful for covering rocks or walls, where they will generally grow with little trouble. The easiest mode of propagating and getting them to grow on such places is first to make the place fit for their reception, by putting thereon a little loam made with a paste of cow-dung; then chopping the plants in small pieces, and strowing them on the place: if this is done in the spring, the places will be well covered in a short time.

45. STATICE Armeria. THRIFT.—This plant is valuable for making edgings to the flower garden. It should be parted, and planted for this purpose either in the months of August and September, or April and May.

46. STIPA pinnata. FEATHER GRASS.—We have few plants of more interest than this; its beautiful feathery

bloom is but little inferior to the plumage of the celebrated Bird of Paradise. It is frequently worn in the head-dress of ladies.

47. SWERTIA perennis. MARSH SWERTIA.—This is a beautiful little plant, and worth the attention of all persons who are fond of flowers that will grow in boggy land. It is a perennial, and of easy culture.

48. TROLLIUS europaeus. GLOBE FLOWER.—This is also a fine plant: when cultivated in a moist soil its beautiful yellow flowers afford a pleasing accompaniment to the flower border and parterre in the spring of the year. It is easily raised by parting its roots.

49. TULIPA sylvestris.—This beautiful flower is also an inhabitant of our flower-gardens; it is called the Sweet-scented Florentine Tulip. It has a delightful scent when in bloom, and is highly worthy the attention of amateurs of flower gardens. It should be planted in September, and will grow in almost any soil or situation.

50. TYPHA latifolia. TYPHA angustifolia. TYPHA minor.—These are all very fine aquatics, and worth a place in all pieces of water; the foliage forms a fine shelter for water-fowl.

51. VIOLA tricolor. HEART'S-EASE.—Is an annual of singular beauty, and forms many pleasing and interesting varieties.

52. VIOLA odorata must not be passed over among our favourite native flowers. This is of all other plants in its kind the most interesting. It forms also several varieties; as Double purple, Double white, and the Neapolitan violet. The latter one is double, of a beautiful light blue colour,

and flowers early; it is rather tender, and requires the protection of a hot-bed frame during winter. It is best cultivated in pots.

53. VINCA minor. LESSER PERIWINKLE.—This is also a beautiful little evergreen, of which the gardeners have several varieties in cultivation; some with double flowers, others with white and red-coloured corols, which form a pleasing diversity in summer.

54. VINCA major. GREAT PERIWINKLE.-I know of no plant of more beauty, when it is properly managed, than this. It is an evergreen of the most pleasing hue, and will cover any low fences or brick-work in a short space of time. The flowers, which are purple, form a pleasing variety in the spring months.

MISCELLANEOUS ARTICLES

53. BETA vulgaris. I have noticed this plant before, both as to its culinary uses and for feeding cattle: but having received a communication from a friend of mine who resides in the interior of Russia, relative to his establishment for extracting sugar from this root, I cannot omit relating it here, as it appears to be an interesting part of agricultural oeconomy.

"I have here two extensive fabrics for the purpose of making sugar from the Red Beet, and we find that it yields us that useful article in great abundance; i. e. from every quarter of the root (eight bushels Winchester measure) I obtain ten pounds weight of good brown sugar; and this when refined produces us four pounds of the finest clarified lump sugar, and the molasses yield good brandy on distillation. This is not all; for while we are now working the article the cows are stall-fed on the refuse from the vats after mashing; and those animals give us milk in abundance, and the butter we are making is equal to any that is made in the summer, when those animals are foraging our best meads."— Dashkoff, in the government of Orel, 1500 miles from St. Petersburgh, Jan 7, 1816.

The above account, which is so extremely flattering, may no doubt lead persons to imagine that the culture of the beet for the same purpose in this country might be found to answer: and as it is our aim in this little work to give the best information on these subjects without prejudice, I shall beg leave to make use of the following observation, which is not my own, but one that was made on this subject by a Russian gentleman, whom I have long had the honour of enumerating among my best friends; and who is not less distinguished for his application both to the arts and oeconomy, than he is for his professional duties, and his readiness at all times to communicate information for the general good.

"The land where the Beet is grown is of an excellent quality, very deep and fertile, and such as will grow any crop for a series of years without manure. Such soils are seldom found in this country but what may be cultivated to more advantage. In such land, and such alone, will this vegetable imbibe a large quantity of the saccharine fluid; for it would be in vain to look for it in such Beet roots as have been grown on poor land made rich by dint of manure.

"It may also be a circumstance worth remarking, that although the sugar thus obtained is very good for common use, it by no means answers the purpose of the confectioner, as it is not fit for preserving; and for this purpose the cane sugar alone is used; so that although great merit may attach to the industry of a person who in times of scarcity can produce such an useful article as sugar from a vegetable so easily grown, yet when cane sugar can be imported at a moderate rate, it will always supersede the use of the other."

56. PYRUS malus. THE APPLE.—This useful fruit, now growing so much to decay in this country, which was once so celebrated for its produce, is grown in great perfection in all the northern provinces of France; and she supplied the London markets with apples this season, for which she was paid upwards of 50,000 l.; and can most likely offer us good cyder on moderate terms.

The French people, ever alive to improvement and invention, having discovered a mode of extracting sugar in considerable quantity from this fruit, I shall transcribe the particulars of it.

On the Preparation of Liquid Sugar from Apples or Pears. By M. DUBUC. (Ann. de Chim. vol. lxviii.)—"Several establishments have been made in the South of France for making sugar from grapes; it is therefore desired to communicate the same advantage to the North of France, as apples and pears will produce sugar whose taste is equally agreeable as that of grapes, and equally cheap.

"Eight quarts of the full ripe juice of the Orange Apples was boiled for a quarter of an hour, and forty grammes of powdered chalk added to it, and the boiling continued for ten minutes longer. The liquor was strained twice through flannel, and afterwards reduced by boiling to one half of its former bulk, and the operation finished by a slow heat until a thick pellicle rose on the surface, and a quart of the syrup weighed two pounds. By this method two pounds one ounce of liquid sugar was obtained, very agreeable in flavour, and which sweetened water very well, and even milk, without curdling it.

"Eight quarts of the juice of apples called Doux levesque, yielded by the same process two pounds twelve ounces of liquid sugar.

"Eight quarts of the juice of the sour apples called Blanc mollet, yielded two pounds ten ounces of good sugar.

"Eight quarts of the juice of the watery apples called Girard, yielded two pounds and a half.

"Twenty-five chilogrammes, or fifty-pounds of the above four apples, yielded nearly fourty-two pounds of juice; which took three ounces of chalk and the white of six eggs, and produced more than six pounds of excellent liquid sugar.

"In order to do without the white of eggs, twenty pounds of the juice of the above apples were saturated with eleven drachms of chalk, and repeatedly strained through flannel, but it was still thick and disagreeable to the taste; twelve drachms of charcoal powder were then added, and the whole boiled for about ten minutes, and then strained through flannel; it was then clear, but higher-coloured than usual; however, it produced very good sugar. Six quarts of apple-juice were also treated with seven drachms of chalk, and one ounce of baker's small-coal previously washed until it no longer coloured the water, with the same effect.

"Eight quarts of apple juice, of several different kinds and in different stages of ripeness, of which one-third was still sour, were saturated with twelve drachms of chalk, and clarified with the whites of six eggs; some malate of lime was deposited in small crystals towards the end, and separated by passing the syrup very hot through the flannel. Very near two pounds of sugar were obtained.

"Ten pounds of bruised apples, similar to the last, were left to macerate for twenty-four hours, and four quarts of the juice were treated with five drachms of chalk and the white of an egg: it yielded one pound six ounces of liquid sugar; so that the maceration had been of service.

"Twenty-four pounds of the pear called Pillage, yielded nine quarts of juice, which required eighteen drachms of chalk and the whites of two eggs, and yielded about twenty-four ounces of sugar, which was less agreeable to the taste than that of ripe apples.

"Six quarts of juice from one part of the above pears, and two of ripe apples, (orange and girard,) treated with eight drachms of chalk and the whites of two eggs, yielded twenty-six ounces of very fine-tasted sugar, superior to the preceding.

"Six quarts of juice, of an equal quantity of apples and pears, treated with ten drachms of chalk and thirteen of prepared charcoal, deposited some malate of lime, and yielded a sugar rather darker than the preceding, but very well tasted.

"Cadet de Vaux says, that apple juice does not curdle milk, and that a small quantity of chalk added to it destroys some part of the saccharine principle. But eight quarts of juice from ripe apples called orange, which was evidently acid, as it curdled milk and reddened infusion of turnsole and that of violet, were treated with four drachms of chalk and the white of an egg: it yielded twenty-two ounces of syrup, between thirty-two and thirty-three degrees of the hydrometer, which did not curdle milk. Another eight quarts of the same juice evaporated to three-fourths of its volume, and strained, yielded twenty-three ounces of clear syrup, which curdled milk, and was browner than that of the neutralized juice, and approached towards treacle in smell and taste. Perhaps the apple called Jean-hure, used by Mr. Cadet, possesses the valuable properties of furnishing good sugar by mere evaporation. It is necessary to observe, that unless the fire is slackened towards the end the syrup goes brown, and acquires the taste and smell of burnt sugar.

"A hundred weight of apples yield about eighty-four pounds of juice, which produce nearly twelve pounds of liquid sugar. Supposing, therefore, the average price of apples to be one franc twenty cents (tenpence) the hundred-weight, and the charge amounts to forty cents (four-pence), good sugar may be prepared for three or four sols (two-pence) per pound [Footnote: A gramme, fifteen grains English.-A drachm, one-eighth of an ounce.]. The only extra apparatus necessary is a couple of copper evaporating pans."−Retrospect, vol. vi. p. 14.

The distressed state of our orchards in the Cider counties has lately much engaged the attention of all persons who are accustomed to travel through them; and no one can possibly view the miserable condition of the trees, without being forcibly struck with their bad appearance: the principal case of which, I am sorry to say, has arisen from mismanagement [Footnote: Vide Observations on Orchards, lately published by the author of this work.]; and it certainly does in a great measure tarnish the laurels of our boasted agriculturists, when we find such great quantities of this useful fruit produced in France, that very country which we have been taught to believe so greatly behind us in the general oeconomy of life.

57. SPERGULA arvensis.—This plant has been recommended as a crop for feeding cattle, and is stated to be cultivated for that purpose in some parts of Germany and Flanders: but I believe we have many other plants better calculated for the purpose here.

58. VIOLA odorata.—This is a very useful plant in medicine, affording a syrup which has long been used in the practice. It is however discarded from the London Pharmacopoeia.

59. URTICA canadensis. CANADIAN HEMP NETTLE.— During the late war, when, from unfortunate circumstances and misunderstandings amongst the potentates of Europe, the commercial intercourse was checked, great speculations were made among the people to discover substitutes for such articles as were of certain demand; and one of the principal was of course the article Hemp, which, although it can be partially cultivated in this country, is a plant of that

nature that we should find the article at a most enormous price were we dependent on our own supply alone. The great growth that supplies all the markets in the world is Russia, where land is not only cheap, but of better quality than here; but with which country we were once unhappily deprived of the advantage of trade. This caused persons to seek for substitutes: and I once saw one that was made from bean-stalks, not to be despised; but it is probable that none has reached so high in perfection as that produced from the plant above named. A person has grown and manufactured this article in Canada, and has exhibited some samples in London, which it is said have obtained the sanction of government, and that the same person is now engaged in growing in North America a considerable quantity of this article. As this, therefore, is a subject of great interest to us as a maritime nation, I shall insert the following account that is given of this plant. I am, however, quite unacquainted with its culture or manufacture, and cannot pledge myself for the accuracy of the detail.

"PERENNIAL HEMP. Cultivation.—Affects wet mellow land, but may be cultivated with advantage on upland black mould or loam, if moist and of middling good quality. Manure will assist the produce. It may be planted from the beginning of October to the latter end of March, in drills about fifteen inches asunder and nine inches distance in the drills.

"Propagation.—Sow the seeds in a bed in the month of March, and transplant the roots next autumn twelvemonth, as above directed; or divide the old roots, which is the quickest way of obtaining a crop.

"Time of Harvesting.—If a fine quality of Hemp is desired, mow the crop when it is in full bloom; but should a greater produce of inferior quality be more desirable, it should stand until the seeds are nearly ripe. It should remain in the field about a week after it is mown, and when sufficiently dry gathered in bundles and stacked as Hemp.

"Separation of Hemp from the Pulps.—Rot it in water, as practised with Hemp.

"The Perennial Hemp grows to the height of from four to six feet.

"The root inclines horizontally with numerous fleshy fibres at the extremity.

"The buds many, and resembling the buds of the Lily of the Valley.

"It is the Urtica canadensis of Kalm, one of which was brought over and planted by the side of this plant, and we could not find any difference."

60. LAPSANA communis. NIPPLE-WORT.—This plant is considered by the country people as a sovereign remedy for the piles. The plant is immersed in boiling water, and the cure is effected by applying the steam arising therefrom to the seat of the disease; and this, with cooling medicine and proper regimen, is seldom known to fail in curing this troublesome disease.

61. DAPHNE laureola. WOOD LAUREL.—The leaves of this plant have little or no smell but a very durable nauseous acrid taste. If taken internally in small doses, as ten or twelve grains, they are said to operate with violence by stool and sometimes by vomit, so as not to be ventured on with safety, unless their virulence be previously abated by long boiling, and even then they are much to precarious to be trusted to. The flowers are of a different nature, being

in taste little other than mucilaginous and sweetish, and of a light pleasant smell. The pulpy part of the berries appears also to be harmless. The bark macerated in water has of late been much employed in France as a topical application to the skin for the purpose of excoriating and exciting a discharge.

62. RUMEX acutus. SHARP-POINTED DOCK.—The root of this plant has long been used in medicine, and considered as useful in habitual costiveness, obstructions of the viscera, and in scorbutic and cutaneous maladies; in which case both external and internal applications have been made of it. A decoction of half or a whole drachm of the dry roots has been considered a dose.—Lewis's Mat. Medica.

63. ELYMUS arenarius. ELYMUS geniculatus. LIME GRASS.—The foliage of these grasses make excellent mats and baskets; and where they grow in quantity afford a livelihood to many industrious persons who manufacture these articles.

64. SALSOLA Kali. GLASS-WORT, or KELP. Soda and Barilla are yielded by this plant. The ashes of this vegetable yield an alkaline salt, which is of considerable use for making glass, soap, &c. The small quantity grown in this country is by no means equal to the demand, and Spain has the advantage of trade in this article, where the plant grows wild in the greatest abundance. An impure alkali similar to these is obtained from the combustion of other marine plants, as the Fuci, &c. by the people in Scotland.

65. BORAGO officinalis. BORAGE—A fine cooling beverage is made from this herb, called Cool Tankard. It is merely an infusion of the leaves and flowers put into water, with the addition of wine, nutmeg, &c. &c.

OBSERVATIONS on the BLEEDING TREES, and procuring the Sap for making Wine, and brewing Ale.

In the article BIRCH TREE, (p. 34, No. 107, of this volume,) we have mentioned the abstracting the sap for the purpose of making wine; and as this is practicable, and may be obtained in some places at little expense and trouble, I shall take the liberty of transcribing the following curious paper on the subject.

"To obtain the greatest store of sap in the shortest time from the body of a tree, bore it quite through the pith, and the very inner rind on the other side, leaving only the bark unpierced on the north-east side. This hole to be made sloping upwards with a large auger, and that under a large arm near the ground. This way the tree will in a short time afford liquor enough to brew with; and with some of these sweet saps, one bushel of malt will make as good ale as four bushels with ordinary water. The Sycamore yields the best brewing sap.

"The change of weather has a great effect on the bleeding of plants. When the weather changes from warm to cold, Birch ceases to bleed, and upon the next warmth begins again: but the contrary obtains in the Walnut-tree, and frequently in the Sycamore, which upon a fit of cold will bleed plentifully, and, as that remits, stop. A morning sun after frost will make the whole bleeding tribe bleed afresh.

"From the latter end of January to the middle of May trees will bleed. Those that run first, are the Poplar, Asp, Abele, Maple, Sycamore. Some, as Willows and the Birch, are best to tap about the middle of the season, and the Walnut towards the latter end of March.

"When a large Walnut will bleed no longer in the body or branches, it will run at the root, and longer on the south or sunny side than on the north or shady side.

"A culinary fire will have the same or greater effect than the sun, and immediately set trees a-bleeding in the severest weather. Branches of Maple or Willow cut off at both ends, will bleed and cease at pleasure again and again as you approach them to or withdraw them from the fire, provided you balance them in your hand, and often invert them to prevent the falling and expence of the sap; but at length they cease.

"A Birch will not bleed however deeply the bark only may be wounded: it is necessary to pierce into the substance of the wood."—Dr. Tonge in Phil. Trans. No. 43.

THE END